Brain-Powered Science

Teaching and Learning With Discrepant Events

Brain-Powered Science

Teaching and Learning With Discrepant Events

Thomas O'Brien

press

National Science Teachers Association

Arlington, VA

National Science Teachers Association

Claire Reinburg, Director
Jennifer Horak, Managing Editor
Andrew Cocke, Senior Editor
Judy Cusick, Senior Editor
Wendy Rubin, Associate Editor
Amy America, Book Acquisitions Coordinator

ART AND DESIGN
Will Thomas Jr., Director, Cover and interior design
Cover and inside illustrations by Daniel Vasconcellos unless noted otherwise.

PRINTING AND PRODUCTION
Catherine Lorrain, Director
Nguyet Tran, Assistant Production Manager

NATIONAL SCIENCE TEACHERS ASSOCIATION
Francis Q. Eberle, PhD, Executive Director
David Beacom, Publisher

Copyright © 2010 by the National Science Teachers Association.
All rights reserved. Printed in the United States of America.
14 13 12 11   5 4 3 2

LIBRARY OF CONGRESS CATALOGING-IN-PUBLICATION DATA
O'Brien, Thomas.
  Brain-powered science : teaching and learning with discrepant events / By Thomas O'Brien.
     p. cm.
  Includes index.
  ISBN 978-1-935155-10-2
  1. Science--Study and teaching (Middle school) 2. Science--Study and teaching (Secondary) 3. Visual educa-
tion. I. Title.
  LB1585.O28 2010
  507.1'2--dc22
                              2009046351

e-ISBN 978-1-936137-75-6

NSTA is committed to publishing material that promotes the best in inquiry-based science education. However, conditions of actual use may vary, and the safety procedures and practices described in this book are intended to serve only as a guide. Additional precautionary measures may be required. NSTA and the authors do not warrant or represent that the procedures and practices in this book meet any safety code or standard of federal, state, or local regulations. NSTA and the authors disclaim any liability for personal injury or damage to property arising out of or relating to the use of this book, including any of the recommendations, instructions, or materials contained therein.

About the Cover—Safety Issues: In the cartoon drawing on the cover, artist Dan Vasconcellos depicts the energy and excitement present in a school science lab. During actual school lab investigations, students should always maintain a safe distance from the teacher who is doing the demonstration (in this case, a two-balloon balancing act). The teacher and students should wear personal protection equipment if the demonstration has any potential for bodily harm. Safety Notes throughout this book spell out when a demonstration requires that the teacher and students wear safety goggles or other protective items.

Contents

Section 1: Introduction to Interactive Teaching and Experiential Learning

Section 2: Human Perception as a Window to Conceptions

Section 3: Nature of Cognition and Cognitive Learning Theory

Contents

Acknowledgments

I owe an immeasurable debt of gratitude to the great science teachers whom I have had the pleasure to learn from and to work with over the years.

My initial inspiration to become a science teacher came from Dan Miller, my high school chemistry and physics teacher and student-teaching mentor. Dan's frequent use of demonstrations and his emphasis on the historical evolution of theories made science both fun and mentally engaging. His gift of the book *Tested Demonstrations in Chemistry*, edited by Hubert Alyea and Fred Dutton and now out of print, catalyzed my interest in exploring the science behind the "magic" of science demonstrations.

When I was an undergraduate student at Thomas More College, the chemistry faculty supported my development of "edu-taining" Chemistry Is Magical programs for elementary classrooms. At the beginning of my work as a secondary science teacher, I was encouraged by Mickey Sarquis and the Cincinnati section of the American Chemical Society (ACS) to develop the skills and confidence to "teach teachers" via the Expert Demonstrator Training Affiliate program.

Later, my mentor at the University of Maryland-College Park, Dr. Henry Heikkinen, guided my dissertation study of a NSF-funded Institute for Chemical Education summer professional development program on chemical demonstrations. Henry's expertise as a writer, editor, and science teacher educator also facilitated my transition to becoming a full-time science teacher educator through early development work on the ACS's Chemistry in the Community textbook. Twenty years later, his insightful critique and encouragement helped me to frame the dual focus of this book: discrepant-event science activities and their use as analogies for science teacher education.

As a professor at Binghamton University, I've benefited from co-teaching grant-funded summer institutes with wonderful colleagues in all four of our science departments. Physicists Andy Telesca (who also reviewed early versions of this book) and Dr. Carl Stannard were especially supportive at the early stage of my development of the dual focus pedagogical strategy. Informal feedback from hundreds of preservice and inservice science teachers has enabled me to refine this approach. I especially appreciate the meticulous review of the science explanations in this book by my former doctoral student, Dr. Douglas Green.

(continued)

Acknowledgments

I would also like to acknowledge the many scientists and science teacher educators whose independent development of discrepant-event demonstrations and analogies is the foundation of my synthesis of these two teaching strategies. Nearly every science activity in this book has a history that goes back to books published at least 60 years ago; a few activities even go back as far as the late 1800s. Isaac Newton's acknowledgment that he "stood on the shoulders of Giants" is especially relevant with my book.

Finally, I would like to thank my wife and children for their encouragement and support. Everyone's children deserve the very best education that we can provide as we continually strive to grow as both teachers and learners.

NATIONAL SCIENCE TEACHERS ASSOCIATION

About the Author

Dr. Thomas O'Brien's 33 years in science education began in K–12 schools, where he taught general, environmental, and physical sciences and high school chemistry. For the last 23 years, he has directed the preservice and inservice, graduate level, science teacher education programs of the School of Education at Binghamton University (State University of New York/SUNY). His master's-level courses include Philosophical and Theoretical Foundations of Science Teaching, Curriculum and Teaching in Science, and Elementary Science Content and Methods. He also supervises the student teaching practica. In addition, he teaches a cross-listed doctoral/postmaster's educational leadership course.

Concurrent with and subsequent to earning a MA and a PhD in Curriculum and Instruction/Science Education at the University of Maryland-College Park, Dr. O'Brien served as a curriculum development specialist and teacher's guide editor on the first edition of the American Chemical Society's *Chemistry in the Community* (*ChemCom*) (1988) textbook and as the co-author of the *New York Science, Technology & Society Education Project Teacher Guide* (1996).

As a science teacher professional development specialist, he has co-taught 20 summer institutes, including national programs of the Institute for Chemical Education and state and regional programs funded by grants from the National Science Foundation, the Howard Hughes Medical Institute, and the New York State Education Department, among others. He has received awards for excellence in teaching and/or service from the American Chemical Society (for National Chemistry Week programs), the New York State Association of Teacher Educators, the SUNY chancellor, and the New York State Science Education Leadership Association. These grants and awards are a reflection of collaborations with university-based colleagues and of what he has learned with and from the large number of K–12 teachers he has had the privilege to serve.

Introduction

As current (or future) grades 5–12 science teachers, professional development specialists, or college-level science teacher educators, you have both the privilege and responsibility of asking your students and colleagues to join you as active, lifelong learners. This book invites you to engage in science that involves both hands-on play and minds-on mental processing. The 33 activities will lead you to critically examine and translate into practice your ever-evolving understanding of science and both the science and the art of science teaching. The "dual-purpose" activities—so called because they address science content and science education—are made up of two components:

1. *Discrepant-event science activities for use both in grades 5–12 classrooms and as models of inquiry-oriented science lessons for use in preservice classes and inservice professional development settings.*

    Whether done as a hands-on activity or demonstration, the discrepant event's surprising, often counterintuitive outcome creates cognitive disequilibrium that temporarily throws learners mentally off-balance. For example nearly everyone "knows" that a sharp needle will pop a balloon, but in Activity #20 learners observe a long, sharp needle skewer a balloon without bursting the balloon. The unexpected outcome of such a discrepant event generates a need-to-know that motivates learners to thoughtfully reconsider their prior conceptions.

    Discrepant-event activities can be used anywhere in a unit. They are especially effective for diagnostic and formative assessment of learners' evolving mix of science conceptions and misconceptions. Teaching science via multisensory experiences with live science phenomena also models the nature of science and contributes to memorable and transferable learning.

    The activities were selected to meet the six criteria of being safe, simple, economical, enjoyable, effective, and relevant for both teachers and their students (see Appendix A for a discussion of the criteria).

2. *"Visual participatory analogies"—that is, visual science education analogies—to catalyze the teacher-as-student's creative use of research-informed science education principles.*

    Teachers commonly use verbal analogies to help students understand new, unfamiliar science concepts in terms of more familiar, better-understood ones (e.g., the cell *is like* a factory). Unfortunately, teachers

Introduction

themselves are not typically provided similar conceptual scaffolds when they become students in science education courses or professional development programs. Visual participatory analogies are a new instructional strategy developed by the author for teaching education theory to teachers. With this strategy, teachers interactively participate with discrepant science phenomena in ways that metaphorically help them bridge the gap between science education theory and practice. For example, Activity #2 uses hands-on play with a Möbius strip as a visual participatory analogy for the interactive nature of teaching and learning.

Your participation as teacher-as-learner-experimenter (rather than simply passive reader) in these minds-on activities will lead you to question, and help you to revise, your implicit assumptions about the nature of science, teaching, and learning. At the same time, you will develop expertise with activities that you can use with your own students. The dual-purpose activities thus allow you to unlock two doors with one key—the doors to your own learning and to your students' learning.

At this point, if you have a burning desire for a direct experience with this science content–science education approach, go directly to a sample activity (e.g., Activity #3, "Burning a Candle at Both Ends: Classrooms as Complex Systems") right now and read the remainder of this Introduction after you have worked and played through the activity. *This book does not need to be used in a strictly linear, front-to-back sequence with your students.* Alternatively, you can read this Introduction (which also describes the book's organizational structure and the activity format), review the related research cited in the Appendixes, and then proceed to activities #1 and #2. These first two activities are introductions to the use of analogies and the idea of interactive teaching and learning that are featured in all subsequent activities.

This book attempts to bridge the gaps between scholarly cognitive science education research, national standards, and teacher-friendly activity books. It asks you to alternate between the roles of student-learner and teacher–reflective practitioner. I hope you will have as much fun with these dual-purpose activities as I have had in developing them during the course of my many years of working with teachers. A second volume is currently under development.

Ways to Use This Book

Preservice Science Methods Courses

This book can be used in preservice science methods course as a supplement to middle and secondary methods textbooks that convey information about constructivist teaching. I believe that every methods class should be a lively, do-as-I-do exemplar of science inquiry approaches. As such, this book's 33 discrepant-event activities (and over 100 extension activities) can be modeled by the instructor and used both in student-presented microteaching lessons and as a resource for fieldwork experiences and student teaching. The science education analogy associated with each activity can be discussed in the methods class and further explored in online forums to emphasize learning as an act of minds-on cognitive construction.

Grades 5–12 Science Classes

Teachers in grades 5–12 science classes can read, practice, adapt, implement, videotape, self-analyze, and further refine the book's model science inquiry lessons. The science content information and science education analogy associated with each activity provide a broad context for the theoretical foundation of minds-on science teaching. Rather than merely being another source of neat activities to add to one's bag of tricks, this book is designed to encourage teachers to critically examine some of their own favorite activities to see how to increase the activities' inquiry potential or how to connect the activities to "big ideas" and scientific habits of mind.

Professional Development for Teachers

This book can also be used in collaborative, teachers-helping-teachers professional development. Whether you are a preservice, novice inservice, or veteran teacher, and whether you majored in science as an undergraduate or not, your own career-long, inquiry-based learning is essential to maintain your professional vitality. Increasingly, state and local school district policies and professional organizations such as the National Science Teachers Association are promoting and supporting continuous professional development (NSTA 2006, 2007a).

In fact, the professional development literature describes a wide variety of models for inservice teacher learning (Banilower et al. 2006; Loucks-Horsley et al. 1998; NRC 2001a; NSTA 2006, 2007a, 2007c; Rhoton and Bowers 2003; Yager 2005). Informal, one-to-one peer collaborations that share the wisdom of practice that resides in any school are too often an untapped resource for professional growth and curricular change. Pairs or small teams

Introduction

of teachers can use the activities in this book as starting points for informal "lesson study" (Stigler and Hiebert 1999). On a larger scale, with financial and logistical support from schools and districts, teachers in an entire science department could work together to refine one another's teaching by visiting one another's classrooms to model and critique lessons.

Other forms of professional development rely more on the leadership of a "master teacher." For instance, districts are increasingly supporting mentor–new teacher pairings and science specialist teachers to lead study groups. Additionally, grant-funded collaborations with scientists and science teacher educators at the college level may provide funding and expertise for an academic year of Saturday Science Seminars, for summer institutes, and for specially targeted graduate-level courses.

Teachers are justifiably skeptical about one-shot workshops run by "outside experts," and research indicates that these workshops rarely result in much more than short-lived motivational boosts. That said, even these quick-fix presentations can serve a catalytic role if they are followed by job-embedded support that helps teachers transfer the lessons learned into their science classrooms.

Other Considerations

If this book is used in a professional development course or program, it is best if the majority of the teacher-learners experience the activity "live" before reading the activity. Inquiry-based science teaching is based on the premise that prematurely giving answers (before engaging the learners with phenomena that raise questions for them to explore) can kill curiosity and limit learning effort and outcomes (NSTA 2004). However, if used in a self-study context, some of the element of surprise will necessarily need to be sacrificed. Even here, individual teacher-learners are encouraged to attempt to answer the questions embedded in the activities—by actually doing the activity—before reading the answers, which are intentionally placed at the end of each activity.

Most activities can be modeled in 10–20 min. when used with teachers as model science inquiry lessons or as science education analogies. With time so limited in most professional development settings, the activities are designed to be easy to set up, execute, and clean up. When used as science inquiry activities with grades 5–12 students, completing and processing the activities could take up to a full class period and would optimally be placed in an integrated instructional unit of related concepts and activities that would extend over a 1–2 week period (e.g., using the 5E Teaching Cycle: Engage, Explore, Explain, Elaborate, Evaluate; see Appendix B for a discussion of this teaching cycle).

Organizational Structure of the Book

This book's 33 interactive, experiential learning activities are clustered into three sections, which are discussed below. Professional development specialists and college-level science teacher educators may wish to use the activities as a framework for either a series of professional development sessions or a more formal course. The major theme of the nature of science, teaching, and learning as informed by cognitive science research runs through all the activities (Aicken 1991; Bybee 2002; Cocking, Mestre, and Brown 2000; Lederman 1992, 1999; McComas 1996; Michaels et al. 2008; NRC 2007; NSTA 2000). The individual activities also can be used as independent stand-alones. Individual science teachers not affiliated with a course or professional development program may wish to use the special Science Content Topics section (pages 361–365) to select activities that match their grades 5–12 instructional scope and sequence. In this case, the science education themes will be encountered on a need-to-know basis in the course of regular classroom teaching.

Section 1. Introduction to Interactive Teaching and Experiential Learning

This short, foundation-setting section (activities #1–#3) introduces analogies as a cognitive tool and instructional strategy for interactive teaching and learning. The three activities use science education analogies to challenge teachers to consider alternative ways of seeing their relationship with learners and to consider the power of inquiry-oriented, curriculum-embedded assessment.

Activity #1 is the only one in the book that is *not* framed around a discrepant-event activity (although teachers may want to adapt the activity to teach their students about the complementary roles and responsibilities of teachers and students). It provides a concrete example and model of how to effectively use analogies to help learners to construct well-articulated understandings and avoid generating misconceptions. References are provided to support teachers' ongoing use of analogies to help students understand nonobservable, abstract, or otherwise conceptually difficult science concepts in terms of more easily visualized, familiar phenomena and processes. Activity #2 introduces the idea of interactive, hands-on explorations (HOE) via a simple paper-and-pencil puzzle that asks learners to predict-observe-explain (POE). Activity #3 demonstrates how guided inquiry can uncover the science behind simple magic tricks.

Introduction

Section 2. Human Perception as a Window to Conceptions

These four activities (#4–#7) each include a number of mini-investigations that encourage learners to playfully explore some of the strengths and limitations of human perceptions (i.e., seeing, hearing, tasting, smelling, and heat flow and pressure-sensitive touching). Humans perceive, process (i.e., reconstruct and conceive), retain, and retrieve only the small portion of external reality that their sensory systems have evolved to detect, based on the selective, adaptive advantages provided (i.e., we notice on a "need-to-notice" basis). Also, to some extent, we perceive what we expect to perceive based on past experiences and preconceptions; human observations are always somewhat theory-laden. As such, our senses can be viewed analogically as tinted or foggy windows that allow small segments of filtered, external reality to enter into human consciousness and form the raw material for our conceptions (and possible misconceptions) about the nature of reality. Understanding our species-specific sensory limitations and individual attention deficits and learning how to design and use technology to help us extend the range, sensitivity, and reliability of our perceptions are central to the nature, history, and ongoing evolution of science that relies on valid, reliable, and "unbiased" empirical evidence.

Section 3. Nature of Cognition and Cognitive Learning Theory

Four major principles of cognitive learning theory are experientially developed through the 26 activities (#8–#33) that make up this major section of the book.

1. Knowledge transmission and passive reception models of teaching and learning are "unquestioned answers" that underlie common schooling practices that overemphasize teaching as telling and learning as listening (Michael and Modell 2003). By contrast, the research-validated idea of learning as a minds-on act of cognitive construction has the power to transform science education. Three hands-on explorations (activities #8–#10) are used to challenge outdated learning theories and provide multisensory experiences that support a more learner-active, constructivist model of understanding (which is further developed in subsequent activities).

2. Learning is a psychologically active, inside-out and outside-in process that is built on two-way interactions between and among individual minds and external learning environments. As such, learning depends on unique intrapersonal factors, interpersonal interactions (i.e., teacher $\leftrightarrow$ learners and learners $\leftrightarrow$ learners), and intentionally designed educational contexts. Effective teaching activates learners' attention and catalyzes cognitive

processing. This general idea is introduced with two activities (#11–#12) and then experientially expanded on in the form of seven approaches that teachers can use to increase their pedagogical powers and instructional effectiveness (14 activities; #13–#26). These seven approaches might be viewed analogically as "weapons of mass instruction" that create pedagogical shock and awe to cause learners to pause, perceive, and ponder:

- Novelty/Changing Stimuli (activities #13–#14)
- Puzzles and Discrepant or Counterintuitive Events (activities #15–#16)
- Cognitive Connections and Meaningfulness (activities #17–#18)
- Multisensory Experiences and Multiple Contexts (activities #19–#20)
- Emotional Engagement, Connections, and Relevance (activities #21–#22)
- Adequate Time for Learning (activities #23–#24)
- Psychological Rewards (Gain/Pain or Benefit/Cost Ratio) (activities #25–#26)

3. Learners' prior knowledge (including preconceptions and/or misconceptions) and cognitive inertia (or "conservatism") may play a constructive, foundational role or a restrictive, limiting role relative to conceptual changes. Just as a solid house cannot be built on a weak foundation, new mental constructions will only stand the test of time if they are built on solid conceptual antecedents. Effective teachers activate and diagnostically assess learners' preinstructional understanding to check for valid precursor ideas, experiential and conceptual holes, and misconceptions. Although many new ideas can be readily assimilated in the context of preexisting ones, new knowledge often requires conceptual accommodation whereby the learners' prior conceptual networks must change for the new information to fit into the picture and make sense (activities #27–#29).

4. Effective science instruction catalyzes cognitive construction and builds a foundation for more independent learning by inviting inquiry rather than by indoctrinating. The last four activities (activities #30–#33) recapitulate the book's major theme of interactive teaching-learning that supports learners' active, minds-on cognitive construction. FUN and MENTAL activities that *engage* learners with discrepant phenomena, raise questions for *exploration* that demand *explanation*, and are rich in possibilities for *elaboration* are a powerful means of achieving this objective.

Introduction

Format Used in Each Activity

Each of the 33 activities has the following format.

Title

This is intended to forecast the science content and science education content foci of the activity.

Expected Outcome

This section is a short description of the setup and expected result of the activity.

Science Concepts

This section briefly discusses the science concepts exhibited by the discrepant event. The author assumes that teachers reading this book are at least somewhat familiar with the underlying science concepts; will develop a deeper understanding through the inquiry questions and answers built into the Procedure and Debriefing sections; and/or can readily obtain additional background information via the Extensions and Internet Connections sections. The activities cut across physical, life, and Earth science concepts with an emphasis on foundational physical science concepts that lend themselves to shorter, mini-experiments and science education analogies. That said, over half of the activities contain a substantive link to biological analogies and applications (see Science Content Topics, pp. 361–365).

Science Education Concepts

When used with teachers (as the targeted learners), each discrepant-event science activity also serves as a visual participatory analogy—or science education analogy—for a science education principle. The intent is to create a common experiential foundation for subsequent collegial conversations and collaborations on the science and art of minds-on science teaching strategies. The long-range goal of the activities is to increase teachers' science content knowledge and pedagogical content knowledge *simultaneously* (Cochran 1997; Shulman 1986).

Having several different activities for each science education principle allows both for instructional flexibility and for key ideas to be introduced, reinforced, and extended in different learning contexts with different analogies. If time permits, experiencing and critiquing multiple analogies for the same science education principle will enable teachers to form a richer, triangulated understanding.* Alternatively, a given activity might be modeled in

---

Triangulate refers to the advantage of using multiple methods or approaches to lead to rich, nuanced answers to a given question. Because any single analogy has its limitations in explaining a given target, when teachers use multiple analogies they help students to develop a more complete understanding of a given scientific concept than they would if only one analogy were used.

a professional development program and teachers could be asked to test-out additional related activities in their classrooms before a second, follow-up session in which they critique and improve the activities.

Materials

This is a list of the required and optional materials needed to complete the activity. Many activities can be done as either an individual hands-on exploration or as teacher or participant-assisted demonstrations, depending on the availability of materials, the time constraints, and instructional setting (i.e., professional development versus grades 5–12 classrooms). Most of the activities use common materials, but *representative* suppliers (and costs) are cited to facilitate easy ordering in cases where unique science equipment or "toys" are used. Although the author has found the cited suppliers to be cost-competitive, no endorsement of particular companies is intended. Additionally, as all prices are subject to change, readers of this book are encouraged to check with the science supply companies used by their local school districts.

Points to Ponder

Each activity includes several powerful quotes from famous scientists, philosophers, or educators. Serious, sustained attention to the history and philosophy of science (HPS) in the K–12 curriculum is called for by research and policy documents (AAAS 1993; Matthews 1994; NRC 1996). Arguments for including more HPS in science courses include the following:

1. Cognitive development (i.e., the idea that a student's cognitive ontogeny at times recapitulates the history of science phylogeny with respect to limited applicability models and misconceptions)
2. The need for a science-and-technology-literate citizenry that understands the nature and evolution of science
3. The benefits of situating and contextualizing science as a human endeavor that both affects and is affected by multicultural, historical forces

Brief historical quotes cannot do justice to HPS, but they can serve as catalysts to teachers to explore other HPS resources (e.g., Asimov 1976; Hakim 2004, 2005, 2007; Hellemans and Bunch 1988; Gribbin 2002; Silver 1998). The discussion questions in the Debriefing sections are explicitly linked to the quotes to raise HPS awareness and interest.

Introduction

Procedure

This section includes the functional description of one or two possible ways of doing the activity. As needed, separate descriptions are provided for two settings: "When Working With Teachers" (i.e., teachers experiencing the activity as professional development or as preservice teachers) and "When Working With Students" (i.e., in grades 5–12 classrooms). The sample inquiry questions typically include attention to "big picture" unifying concepts or themes (drawn from the National Science Education Standards [NRC 1996] and the Benchmarks for Science Literacy [AAAS 1993]) that guide the learners to use empirical evidence, logical argument, and skeptical review to make and revise hypotheses about what is occurring and why. Meaningful learning occurs when teachers build on knowledge- and comprehension-type questions (e.g., What do you observe?) up to questions that require higher-order thinking skills associated with application, analysis, synthesis, and evaluation (e.g., How do you account for and apply the science underlying your observations?) (Anderson and Krathwohl 2001; Bloom et al. 1956).

The sample inquiry questions in this section and the Debriefing section are not intended to be used verbatim; rather, they suggest possible productive lines of inquiry and model the art of effective science questioning.

> Effective questioning that elicits quality responses is not easy. In addition to optimal wait time, it requires a solid understanding of subject matter, attentive consideration of each student's remarks, as well as skillful crafting of further leading questions. (NRC 2001b, p. 35)

Questions posed by the teacher serve multiple pedagogical purposes. They catalyze two-way teacher-student interactions that go beyond a simple sequence of teacher question (initiation) → student response → teacher feedback that serve to "move a lesson along." They also provide formative assessment to determine if students are "getting it" and the opportunity for clarifications and deeper probing of student conceptions. More important, teacher-initiated questions explicitly model for students how to ask their own scientifically productive questions that lead to fruitful, inquiry-based examination of phenomena by students and interactions among students. As such, the questions generated by students provide a window into their cognitive processing and evolving conceptions, perhaps even more so than their answers to teacher questions.

Additionally, student-initiated questions help students learn important metacognitive skills related to learning how to learn and to developing the intellectual dispositions and habits of mind of active, engaged learners. Together, teacher- and student-initiated questions create a collaborative classroom environment based on a shared dialogue of discovery.

Debriefing

This section describes some of the broader context and lessons-to-be-learned about the science education and the science content. As in the Procedure section, separate "When Working With Teachers" comments (focused on science pedagogical knowledge) and "When Working With Students" comments (focused on science content) are provided as needed. The comments may also provide additional tips for teachers when using the activity to teach science to their grades 5–12 students. If desired, the teacher debriefing questions can be used as "homework" and/or discussed via electronic learning communities in live professional development sessions (NSTA 2008).

Extensions

These are brief descriptions of related "what if I were to change…" activities for further exploration as time and interest allow. Given the limited time in professional development settings, the extensions are especially useful for independent self-assessment work by teachers to determine if they really "get it." The extensions also provide complementary activities that could be used to help design 5E Teaching Cycles or integrated instructional units for grades 5–12 science instruction (see Appendix B for a description of the 5E cycle). Also, when the science content connects with another activity in the book, the related activity is cited. *The Extensions increase the number of distinct science inquiry activities in this book to to nearly 120.*

Internet Connections

In this list, readers will find up-to-date links to a variety of supplemental web-based resources including the following:

- Video clips of similar or related demonstrations where teachers can watch another teacher perform the mechanics of the demonstration
- Animations and interactive simulations that teachers can use to help the students visualize science principles and processes that are at scales that are too small/large, too fast/slow, or too dangerous or expensive to be seen with the unaided eye or realistically manipulated by students. Some of these websites (e.g., *http://phet.colorado.edu*) contain extensive libraries of simulations and related teaching materials that cut across science disciplines.

Introduction

- Online encyclopedias that further explain the science content and related real-world applications
- Short professional development readings related to the science education analogy

E-learning experiences and resources represent an ever-growing venue for teacher professional development and "just-in-time" instructional resources for teaching K–16 science (NSTA 2008). The internet is in a continual state of flux, but the majority of the cited web pages originate from relatively stable, nonprofit organizations (e.g., museums, professional associations, and universities). In addition to these websites, each of which has been reviewed by the author for relevance to the activities in this book, teachers may explore the science content and related curricular materials more broadly through NSTA SciLinks (*www.scilinks.org*).

In either case, occasional encounters with "dead" links are the equivalent of a book or journal going out-of-print or otherwise becoming unavailable—*except* that in the case of the internet, other great resources are always beckoning a few keystrokes away. As such, the sites provided should be viewed as starting points for further explorations. In addition to their inclusion in the book, an NSTA Press online, hyperlinked resource will allow readers to electronically access the sites in spring 2010.

Answers to Procedure and Debriefing Questions
The answers to the questions in the Procedure and Debriefing sections are deliberately presented at the very end of each activity. This was done to maintain the inquiry nature of the book. Attempting to answering the questions in the context of doing the activity (rather than reading the answers first) will help the teacher enjoy the activity more, appreciate the challenge that inquiry questions pose for his or her own students, and improve the teacher's own questions and answers.

Conclusion

As teachers, we tend to teach both what and how we were taught during our "apprenticeships of observation" as K–16 students (Lortie 1975). It's great to be able to stand on the shoulders of our own exemplary, former science teachers, but research on how to facilitate learning is always advancing. As such, this book challenges you to "question the answers" of your own past experiences as students and to make a paradigm shift away from any pedagogical beliefs and practices that no longer make sense in the light of today's research-informed standards.

The National Science Teachers Association has long recognized reflection-in-action by "teacher action-researchers… [as the] basis for curricular and instructional reform" (NSTA 1990; see also Schön 1983). The authors of the National Science Education Standards (NRC 1996) concur:

> The vision of science and how it is learned will be nearly impossible to convey to students in schools if the teachers themselves have never experienced it…preservice programs and professional development activities for practicing teachers must model good science teaching (p. 56)…. Involve teachers in actively investigating phenomena that can be studied scientifically, interpreting results, and making sense of findings consistent with currently accepted scientific understanding (p. 59)…. Teachers also must have opportunities to engage in analysis of the individual components of pedagogical content knowledge—science, learning, and pedagogy—and make connections between them. (p. 63)

This book's combined science content–science education focus is designed to help current (or future) grades 5–12 science teachers, professional development specialists, and college-level science teacher educators achieve this standard.

The teacher is the key to change and learning in the classroom (NCMST 2000; NCTAF 1996, 1997; NSB 2006; NSTA 2007a). In fact, "the single most important factor affecting student academic gain is teacher effect" (Sanders and Rivers 1996). Some science teachers mistakenly believe that factors outside their control—such as family income, parent education levels, and race or ethnicity—are acceptable explanations for many of their students failing to learn science. On the contrary, effective teachers can cumulatively have a greater impact on educational outcomes than those factors (Ferguson and Ladd 1996). Specifically, the use of engaging activities in every science class is an example of something that *is* in the teacher's control as is teacher collaboration in continuous professional development.

The book is the result of mutually beneficial interactions I have had with hundreds of dedicated science teachers over the last 30 years (e.g., O'Brien 1992a, 1992b; Stamp and O'Brien 2005; Stannard, O'Brien, and Telesca 1994). Please use, improve, and share these activities with your colleagues and students. I hope that you find this book to be "edu-taining" in ways that extend well beyond the initial surprise value and motivational impact of the individual activities. The best teaching and learning experiences are about sharing, catalyzing change in others, and being changed in the process.

Science Education Topics

This book has two focuses—science education and science concepts. The author has designed two alternative tables of content—in addition to the traditional one on pages v–vi—that are organized by these two focuses. The table of contents that begins on this page is organized by science education topics; the table of contents organized by science concepts begins on page 361.

Acronyms Used in Science Education Topics

BBS: Black Box System: A hidden mechanism is explored via observation and testable inferences.

BIO: Biological analogies and applications are specifically highlighted.

HOE: Hands-On Exploration: Learners working alone or in groups directly manipulate materials.

MIX: Mixer: Learners assemble themselves into small groups based on a specific task.

NOS: Nature Of Science: These activities focus on empirical evidence, logical argument, and skeptical review.

PAD: Participant-Assisted Demonstration: One or more learners physically assist the teacher.

POE: Predict-Observe-Explain: The activities use an inquiry-based instructional sequence.

PPP: Paper and Pencil Puzzle: The activities use a puzzle, which is typically focused on the NOS; often a BBS.

STS: Science-Technology-Society: The focus is on practical, real-world applications, and societal issues

TD: Teacher Demonstration: The teacher manipulates a system and asks and invites inquiry questions.

TOYS: Terrific Observations and Yearnings for Science: The activity uses a toy to teach science.

Science Education Topics

Section 1. Introduction to Interactive Teaching and Experiential Learning

Activity	Activity Type	Science Concepts
1. Analogies: Powerful Teaching-Learning Tools	MIX/PPP p. 3	analogies as conceptual tools (*This is the only activity that is not a science discrepant event.*)
2. Möbius Strip: Connecting Teaching and Learning	HOE/PPP p. 15	NOS, POE, topology
3. Burning a Candle at Both Ends: Classrooms as Complex Systems	TD p. 25	POE, phase change, combustion, convection, density, cellular respiration (Extension #3: BIO)

Section 2. Human Perception as a Window to Conceptions

4. Perceptual Paradoxes: Multisensory Science and Measurement	PAD p. 37	sensory adaptations and survival (BIO), (mis)perception, cognition, temperature sensitivity, taste (as related to smell), weight versus density
5. Optical Illusions: Seeing and Cognitive Construction	PPP p. 47	sensory (mis)perception, cognition (BIO); quantitative measurements
6. Utensil Music: Teaching Sound Science	HOE p. 63	sound transmission, perception, sensory variations in species (BIO)
7. Identification Detectives: Sounds and Smells of Science	HOE/MIX p. 73	BBS, NOS, sensory adaptations, survival (BIO), identification by sound, identification by smell

Section 3. Nature of Cognition and Cognitive Learning Theory

Knowledge Transmission and Reception Versus Construction of Understanding

8. Two-Balloon Balancing Act: Constructivist Teaching	HOE or PAD p. 87	NOS, POE, LaPlace's law and surface tension, air pressure, BIOmedical applications (Extension #2)
9. Batteries and Bulbs: Teaching Is More Than Telling	HOE p. 97	complete or closed electric circuits, energy conversions
10. Talking Tapes: Beyond Hearing to Understanding	HOE p. 109	TOYS, sound, information encoding and gene expression, form/function relationships (BIO)

Activity	Activity Type	Science Concepts

Learning as a Psychologically Active, Inside-Out, and Outside-In Process

Activity	Activity Type	Science Concepts
11. Super-Absorbent Polymers: Minds-on Learning and Brain "Growth"	HOE or PAD p. 119	measurement, polymers, TOYS, BIO/evolution, STS tradeoffs, perspiration (Extensions #2 and #4)
12. Mental Puzzles, Memory, and Mnemonics: Seeking Patterns	PPP p. 131	NOS, pattern recognition, cognition (BIO)

*Novelty and Changing Stimuli**

Activity	Activity Type	Science Concepts
13. Sound Tube Toys: The Importance of Varying Stimuli	HOE or PAD p. 141	sound energy, pitch, Bernoulli's principle, TOYS, POE, animal BIOadaptation of noticing novelty
14. Convection: Conceptual Change Teaching	PAD p. 153	heat, equilibrium, density, convection, POE

*Puzzles and Discrepant or Counterintuitive Events**

Activity	Activity Type	Science Concepts
15. Brain-Powered Lightbulb: Knowledge Transmission?	PAD p. 163	complete or closed electric circuit, BIOfuels analogy (Extension #1), TOYS
16. Air Mass Matters: Creating a Need-to-Know	TD p. 171	air pressure, inertia, POE

*Cognitive Connections and Meaningfulness**

Activity	Activity Type	Science Concepts
17. 3D Magnetic Fields: Making Meaningful Connections	TD p. 179	magnetism, force field lines, neural networks, MRI (BIO/Extension #1)
18. Electric Generators: Connecting With Students	PAD p. 189	electric generators ←→ motors, electric circuits

*Multisensory Experiences and Multiple Contexts**

Activity	Activity Type	Science Concepts
19. Static Electricity: Charging Up Two-by-Four Teaching	PAD p. 201	static electricity (triboelectricity)
20. Needle Through the Balloon: Skewering Misconceptions	HOE or PAD p. 211	polymer elasticity, cell membrane model (BIO/Extension #1)

*Emotional Engagement, Connections, and Relevance**

Activity	Activity Type	Science Concepts
21. Happy and Sad Bouncing Balls: Student Diversity Matters	HOE or PAD p. 221	TOYS, POE, potential→kinetic conversion, law of conservation of energy, friction, elasticity, form/function fitness (BIO)

(continued)

*Each of the categories with an asterisk is one of the Seven Principles for Activating Attention and Catalyzing Cognitive Processing (Activities #13–#26). The seven principles have been identified by the author.

Science Education Topics

(continued)

Activity	Activity Type	Science Concepts
22. Electrical Circuits: Promoting Learning Communities	HOE or PAD p. 233	complete or closed electric circuits, energy conversions, TOYS

*Adequate Time for Learning**

Activity	Activity Type	Science Concepts
23. Eddy Currents: Learning Takes Time	PAD p. 241	electromagnetism, Lenz's law
24. Cognitive Inertia: Seeking Conceptual Change	TD/PAD p. 251	Inertia and cognitive conservatism, independence of vertical and horizontal forces and motions

*Psychological Rewards (Gain/Pain or Benefit/Cost Ratios)**

Activity	Activity Type	Science Concepts
25. Optics and Mirrors: Challenging Learners' Illusions	PAD p. 259	optical illusions, mirrors, BBS, NOS, TOYS
26. Polarizing Filters: Examining Our Conceptual Filters	TD p. 267	light polarization, UV protection for skin and eyes (BIO/Extension #1)

Role of Prior Knowledge, Misconceptions, and Cognitive Inertia

Activity	Activity Type	Science Concepts
27. Invisible Gases Matter: Knowledge Pours Poorly	PAD p. 275	gases occupy space (volume)
28. The Stroop Effect: The Persistent Power of Prior Knowledge	PAD/PPP p. 285	NOS, human perception, cognition (BIO)
29. Rattlebacks: Prior Beliefs and Models for Eggciting Science	HOE or TD p. 293	BBS, NOS, TOYS, energy conversion, rotational inertia, model of the lithosphere

Science Instruction Catalyzes Cognitive Construction

Activity	Activity Type	Science Concepts
30. Tornado in a Bottle: The Vortex of Teaching and Learning	PAD p. 301	gases occupy space, POE, TOYS
31. Floating and Sinking: Raising FUNdaMENTAL Questions	HOE p. 309	density/buoyancy, diffusion, osmosis (BIO), nucleation sites, solubility of gases in liquids, NOS, POE
32. Cartesian Diver: A Transparent But Deceptive "Black Box"	HOE p. 321	Archimedes and Pascal's principles, Boyle's law, density/buoyancy, BBS, NOS
33. Crystal Heat: Catalyzing Cognitive Construction	TD/HOE p. 331	phase changes, latent heat, law of conservation of energy. BIO: cellular respiration (Extension #2), perspiration and thermoregulation (Extension #4), bee colony collapse disorder (Extension #5)

---

*Each of the categories with an asterisk is one of the Seven Principles for Activating Attention and Catalyzing Cognitive Processing (Activities #13–#26). The seven principles have been identified by the author.

Section 1:
Introduction to Interactive Teaching and Experiential Learning

Activity 1

Analogies: Powerful Teaching-Learning Tools

Expected Outcome

Teachers explore how teaching shares some attributes with a variety of other occupations; students consider their respective roles as learners; and both consider the reciprocal, interactive nature of the teaching-learning partnership by way of analogies (see pp. xi–xii for a discussion of analogies). This mixer activity is designed for a group-learning context, though individual teachers can use it as a self-reflection activity.

Science Concepts

Other than modeling the science process skill of analogical reasoning, this activity is *not* intended to teach any specific science concept. *This is the only activity in this book that is not based on a science discrepant-event activity that can be used with grades 5–12 students (although teachers may wish to use a modified form of the activity with their students to introduce the idea of analogies and the need for them to be active learners).* Analogies help explain nonobservable, abstract, or otherwise conceptually difficult concepts in terms of more easily visualized, familiar phenomena and processes. Historically analogies have played a critical role as provisional mental models to help scientists move beyond previous, more limited paradigms (e.g., consider the historical evolution of increasingly sophisticated atomic models from the indivisible, uniform density, billiard ball atoms → plum pudding → solar system → quantum mechanical wave/particle duality—all models for what we thought atoms "looked like"). The popular writings of modern-day scientists-authors-teachers such as Isaac Asimov, Richard Feynman, Stephen Jay Gould, Stephen Hawking, Carl Sagan, Lewis Thomas, and Edward O. Wilson are also replete with analogies (e.g., "think of it like this," "this is similar to that because," and "picture it like").

Science Education Concepts

This activity models the use of analogies as both a probe to activate and assess student understanding and a concept-building tool to extend students' prior knowledge. In this activity, the focus is on the interactive, reciprocal nature of teaching and learning and the related implications for curriculum, instruction, and assessment.

Analogies provide a type of cognitive scaffolding that builds bridges between familiar, known concepts (*analogs*) and unfamiliar or unknown ones (*targets*). This is done by explicitly constructing links between the concepts' *shared attributes* (i.e., structural and/or functional properties). This process of constructing links is called mapping. Analogies can be used as internally visualized, as externally verbalized constructs, as external visual images, and/or as actual physical models.

Textbooks commonly use a limited number of analogies (e.g., cell: city; electrical current: water flow in pipes; atoms: billiard balls), typically without highlighting the dissimilarities (or *unshared attributes*) between the analog and target. This pedagogical omission, along with the selection of analogs that are not well understood by students or are not well articulated in the lesson, can actually create misconceptions and leave students with "one more meaningless thing to memorize." Conversely, well-selected analogies are a powerful pedagogical tool that can help students visualize and comprehend abstract concepts and develop the metacognitive skills and dispositions to generate their own analogies. Whether teacher- or student-generated, analogies can motivate students to recall, retrieve, report, refine, and/or reconstruct their prior conceptions in terms of more scientifically accurate, explanatory models (i.e., they learn to think more like scientists).

Materials

- 1, 5 in. × 8 in. index card for each learner
  - For Method A or Method B in Procedure step #2: Prepare the index cards ahead of time by using the table "Sample Analogies for the Interactive Nature of the Teaching-Learning Process," page 11. Select one occupation from the left-hand column and write that word on each of the cards (use from three to five cards). Then select a second occupation and write the word on the next set of three to five cards and continue this until you have enough card sets for your overall group size. A complete class set might consist of four or five different occupation analogy cards.
  - For Method C in Procedure, step #2, index cards are not needed.

image by RypeArts for iStockphoto

Points to Ponder

If you give a man a fish, he will have a meal. If you teach him to fish, he will have a living.

If you are thinking a year ahead, sow a seed. If you are thinking ten years ahead, plant a tree.

If you are thinking one hundred years ahead, educate the people.

By sowing a seed once, you will harvest once. By planting a tree, you will harvest tenfold.

By educating the people, you will harvest one hundredfold.

—Kuan-tzu, Chinese philosopher (4th–3rd century BC)

All genuine learning is active, not passive. It involves the use of the mind, not just the memory. It is the process of discovery in which the student is the main agent, not the teacher…. Teachers may think they are stuffing minds, but all they are ever affecting is the memory. Nothing can be forced into anyone's mind except by brainwashing, which is the very opposite of genuine teaching.

—Mortimer Adler, American philosopher-educator (1902–2001) in *The Paideia Proposal: An Educational Manifesto* (1982)

The first objective of an act of learning over and beyond the pleasure it may give, is that it should serve us in the future. Learning should not only take us somewhere, it should allow us to later go further more easily.

—Jerome Bruner, American cognitive psychologist-educator (1915–)

Procedure

(See answers to questions in step #1 on p. 13.)

1. Begin the activity by sharing the Kuan-tzu quote. Note the use of multiple analogies in the quote. Ask the following focus question: How can the use of analogies help us understand the interactive, experiential nature of the teaching and learning dynamic? Challenge the learners to consider the implications of the sample analogy that will be used in this activity—**teaching as gardening** (or cultivating)—by raising the following questions with the whole class:

   a. In what ways is the work of a teacher (*target* of the analogy) like the work of a gardener (*analog* of the analogy)?
   b. If the teacher is like a gardener, then what is the learner in this analogy?
   c. Is the seed or learner's role more passive or active? Explain.
   d. What are two things the teacher-gardener should know before planting the seeds/working with learners?
   e. Do different types of seeds grow equally well in all environments?
   f. What is the teaching equivalent of, when gardening, using fertilizer and making the best use of water and sunlight?
   g. What is the teaching equivalent of pruning?
   h. What is the teaching equivalent of weeds and rocks?
   i. What are the limitations of this extended analogy? In other words, how is teaching *not like* gardening? How is learning *not like* seed development?

2. Introduce the Adler quote. Discuss how teaching is not "stuffing minds" or "brainwashing." Then choose Method A, B, or C to proceed.

   Method A: Randomly assign the learners to teams of three to five (using a count-off method). Give each team a set of occupation cards (all cards in a set are the same) that you have previously prepared (see Materials).

   Method B: Randomly distribute the prepared occupation cards and have the learners introduce themselves to each other in

order to self-assemble teams that have the same occupation analogy. This approach is useful at the start of a professional development session if you want to get teachers to introduce themselves to more than just the teachers in their small groups (as in Method A).

Method C: Either randomly assign teachers to teams of three to five members or let them self-select their own teams. Then have each team meet and quickly brainstorm possible analogies (without prompting with the sample analogies in the table on p. 11) and select one analogy of their own choosing to develop, analyze, and share with the broader group. This is the least directive, most open-ended of the methods.

Then, ask the teams (again, made up of people with the same card) to discuss their analog cards in terms of the following:

a. *Mapping.* Mapping is the process of comparing shared attributes between target and analog. This step is critical when using analogies with any age learner. In this case, ask: How is teaching (the target of the analogy) like this other occupation (the analog of the analogy)?

b. *Prediction and Application.* If the teacher is like (X occupation) then the learner would be like (Y role).

c. *Extension of the Analogy.* Ask: What does this teacher-learner analogy suggest are the purposes of the teacher's integrated curriculum-instruction-assessment plans? What should these plans do to, for, or with the learners? (This extension activity is important grounding for teachers on the use of analogies. Not every teacher will want to use this extension, or even Activity #1 itself, with his or her students; however, the author has successfully used this extension with ninth graders as an extended analogy of teacher-as-coach and students-as-players as related to daily practice and formative and summative assessment.)

d. *Limitations of the Analogies.* Ask: How is teaching unlike the occupations listed in column one? Where along the continuum of brainwashing–conditioning–training–informing–instructing– inspiring (inspiring here means catalyzing independent, creative thoughts, words, and actions) is the act of teaching?

3. *When Working With Teachers.* Have the different teams share their discussions of their analogies with the whole class and combine the class results to construct a summary table that is similar in form and content to the sample table on page 11. (You may or may not wish to distribute a copy of this sample table at the conclusion of the discussion.) The table created by the teachers can be used to discuss open-ended questions such as the following:

   a. What makes some analogies better than others? What role might humor play in painting memorable visual images?
   b. Which occupations suggest the most active, minds-on roles for the learners?
   c. Do any of the more "far out" analogies open up new insights into some aspects of the complex, multifaceted profession of teaching and the challenges of facilitating learning?
   d. How does the use of multiple analogies (versus a single one) allow us to triangulate and get a better sense of the synergistic, iterative roles played by curriculum, instruction, and assessment?

4. *When Working With Students.* Ask them to focus on their reciprocal roles and responsibilities as learners in relation to the teacher. Grades 7–12 teachers may want to use a modified version of this activity at the start of the year to playfully emphasize the need for students to actively participate in class and to provide feedback to the teacher when they do and do not understand something in a lesson. This activity also can be used as an introduction to the yearlong use of science analogies as a learning tool.

Debriefing

When Working With Teachers

Point out that throughout history, analogical thinking has enabled scientists to bridge the gap between the known and the unknown (the "unknown" being nonobservable or inaccessible entities or processes) and/or allowed them to look at the known from a new angle or perspective. Analogies (and related concepts such as metaphors

and similes) are also major concept-construction and mental model-building tools used by learners anytime they encounter something new that shares attributes with something with which they are already familiar.

Share the Bruner quote with the teachers (and its implied analogy of "learning as traveling") and discuss whether they feel that their experiences with analogical reasoning in this activity "not only take[s] us somewhere, [but allows] us to later go further more easily" (i.e., with respect to using other analogies in studying about the nature of science, teaching, and learning). What assumptions or "unquestioned answers" about the teaching-learning process does this activity challenge? How can teachers and learners take active roles that make each other's work more effective? Share the list of books that feature science analogies for teachers to use (see Extensions, step #2).

The Teaching-With-Analogies model (Glynn 1991) recommends a six-step process when using analogies to teach science concepts:

1. Introduce the target concept
2. Review with students what they know about the analog concept
3. Identify relevant features of the target and analog
4. Map similarities
5. Discuss where the analogy breaks down
6. Draw conclusions

This model and others like it (e.g., Harrison and Coll 2008) emphasize the importance of highlighting the *unshared* attributes or the limitations of the analogy. If the teacher omits this critical step, analogies can create alternative misconceptions in learners that run counter to the teacher's intentions and learning objectives.

Sample Analogies for the Interactive Nature of the Teaching-Learning Process

The Teacher Is Like an/a...	Learners Are Like...	The Teacher Designs Curriculum, Instruction, and Assessment to Carry Out the Role in Column One (at left)
Artist	Potential Art Lovers	*How is the teacher like an artist?* Entertains and creates an appreciation for the artwork and alternative ways of "seeing" and experiencing reality
Coach	Athletes	*How is the teacher like a coach?* Models desired behaviors, motivates practice to develop skills, provides corrective feedback, and strategizes to maximize both individual and team performance and autonomy over time
Conductor	Musicians	*How is the teacher like a conductor?* Orchestrates individual and group interpretations to create shared meaning and develop skills to recreate and perform a score of music for a broader audience
Doctor	Patients	*How is the teacher like a doctor?* Intervenes to assist the immune system to counter disease and promote natural growth and development with both individualized and communal diagnoses and preventive and ameliorative treatments
Engineer	Client with Problem	*How is the teacher like an engineer?* Works within constraints to optimize or fix a system until further challenges require additional modification
Garbage Collector	Home Owners	*How is the teacher like a garbage collector?* Removes nonuseful items (misconceptions) to make room for new, improved products in the homes (minds)
Lawyer	Jury	*How is the teacher like a lawyer?* Convinces jury of the client's claim using empirical evidence, logical argument, and skeptical review (and "showmanship")
Magician	Audience	*How is the teacher like a magician?* Suspends disbelief, builds a sense of wonder, and creates a need-to-know and desire to come back for more
Plumber	Fluid system	*How is the teacher like a plumber?* Unclogs blockages to allow flow through pipes and/or constructs more efficient systems to optimize flow
Salesperson	Consumers	*How is the teacher like a salesperson?* Generates interest in, experience with, and ultimate ownership and skilled use of an appealing product
Translator or Interpreter	Non-native speakers	*How is the teacher like a translator or interpreter?* Mediates cross-cultural, accurate dialogue and meaning-making (including foreign terms for foreign things)
Travel Guide	Travelers	*How is the teacher like a travel guide?* Helps interpret or read a map relative to signs or checkpoints in the real world in order to facilitate and enrich journeying to new and exciting destinations

Extensions

1. *Visual Art Analogies.* The Official M.C. Escher Website (*www. mcescher.com*) sells a variety of posters, T-shirts, and other products based on this artist's unique perspective. Escher's works entitled *Hand with Globe, Reptiles,* and *Tower of Babel* make for great visual analogies or discussion props when working with teachers. They can be used to help teachers focus on reflective practice; the need for continuous feedback loops between curriculum, instruction, and assessment; and the challenge of teaching in ways that truly facilitate learning. These art works can also be used with students to help them see that they must actively construct meaning by connecting new experiences with previous understandings, by assimilating new ideas, and sometimes by adjusting their previous conceptions.

2. *Science Analogy Pedagogical Resource.* Professional development settings are good places to introduce teachers to analogies. See Camp and Clement 1994; Dagher 1998; Duit 1991; Gilbert and Watt Ireton 2003; Glynn 1991; Hackney and Wandersee 2002; Harrison and Coll 2008; Harrison, Coll, and Treagust 1994; Hoagland and Dodson 1995; Lawson 1993; and Packard 1994, as well as the following Internet sites.

Internet Connections

- Analogical transfer—Interest is just as important as conceptual potential (article by A. Harrison): *www.aare.edu.au/02pap/har02431.htm*

- I never metaphor I didn't like: A comprehensive compilation of history's greatest analogies, metaphors, and similes (book of quotations and wordplay website): *www.drmardy.com*

- Instructional Design Models (bibliographies and links on important philosophers and scientists): *http://carbon.cudenver.edu/~mryder/itc/idmodels.html*

- Northwestern University, Cognitive Psychology (pdf files of Dedre Gentner's research on analogies: *www.psych.northwestern.edu/psych/people/faculty/gentner/publications2.htm*

- Teaching Teachers to Use Analogies (a site by Mark James, Northern Arizona University): *www.physics.nau.edu/~james/ TeachingTeachersAnalogies.htm*
- Teaching With Analogies Model (pdf articles by Shawn M. Glynn, University of Georgia): *www.coe. uga.edu/twa*
- UCLA Reasoning Laboratory (pdf files of Keith Holyoak's publications on analogies): *http://reasoninglab.psych.ucla.edu/Keith Publications.htm*
- Wikipedia: Analogy: *http://en.wikipedia.org/wiki/Analogy*

Answers to Questions in Procedure, step #1

The Gardener [Teacher] works to support the natural development of the seed [Learner], which does the actual growing based on its innate potential and responsiveness to various environmental factors. The wise gardener chooses and assesses what kinds of seeds to plant and where and how to plant them relative to the nature and composition of the soil [curriculum scope and sequence] and the needs of those particular kinds of seeds.

A one-size-fits-all environment is not optimal for seeds. Different kinds of seeds will have somewhat different needs for fertilizer, water, and sunlight. Similarly, learners need differentiated instructional plans that use curriculum-embedded diagnostic and formative assessments to periodically "test the soil and measure the plant's growth with the teacher," making adjustments as needed, rather than simply waiting until harvest time. The teaching equivalent of weeds and rocks are in-school and out-of-school environmental factors that divert the attention of the learner, steal resources from the learner, or otherwise act as barriers to be removed, worked around, or overcome.

Limitations of this analogy include the fact that the teacher does not get to select which seeds/learners he or she will plant; does not have control of relevant environmental factors or even the location of the garden plot/school; and may lack resources needed to maximize the growth of all of his or her diverse learners. Most important, a classroom is not a monoculture environment! Unlike seeds,

individual learners (who are diverse with respect to many factors including gender, ethnicity, socioeconomic status, religious affiliation, multiple intelligence profile, and motivation) can consciously decide how to react to, modify, and decide whether to stay in or leave (physically or psychologically) particular environments. Intentional human learning is more creatively adaptive than the growth of an unconscious, stationary plant. As a result, the work of accurately assessing and cultivating *human* development is much more difficult than assessing and cultivating plant development.

Activity 2

Möbius Strip: Connecting Teaching and Learning

Expected Outcome

A thin strip of paper is twisted 180 degrees and taped end-to-end to form a strip that surprisingly has only one continuous surface (rather than two sides). The strip's length is double that of one side of the original strip.

Science Concepts

A Möbius strip is a nonorientable, two-dimensional surface with only one side. The one-sided nature of the Möbius strip is an example of an *emergent property* (i.e., a property that is found in a system as a whole, but not in any part of the system). The Möbius strip serves as discrepant paper-and-pencil puzzle entry into scientific inquiry ("what will happen if..."). It also models the nature of science (NOS) as an empirically based but creatively driven process. The Möbius strip can be used as an introduction to topology—the mathematical study of properties that are preserved through deformations, twisting, and stretching of objects.

Science Education Concepts

Teaching is an interactive process designed to activate attention and catalyze cognitive processing in one or more learners. While we can't "learn a student," we can teach a student and become a learner ourselves at the same time. This topological puzzle serves as a *visual participatory analogy* (go to p. xi for a discussion of this term) to challenge teachers to reconsider the connection among their curricular, instructional, and assessment plans; implementation efforts; and impact on learners.

This puzzle also introduces the predict-observe-explain (POE) approach to teaching with discrepant events (*discrepant events* is discussed on p. xi. Activities #2–#33 in this book are all based on discrepant events). Students "playing" with this simple, hands-on-exploration can be challenged to think about how learning requires minds-on effort on their part. They need to exert this mental effort if they are to interact with the phenomena and ideas introduced by their teacher. Both teacher and students should see the need to stay connected with—and provide feedback to inform and to teach—each other.

Materials

- Transparent tape
- Scissors
- A 1 in. wide × 14 in. long strip of paper for each learner. Prepare these strips by photocopying on one side the following (including

arrow line) TEACHING --
----------------------------→

- On the reverse side, photocopy upside down but with the arrow pointing in the same direction (when viewed through the topside to the backside) LEARNING ---
----------------------------------→

- *Note:* A font size of 50–60 pts. (Times Roman) will fit on a 1 in. long strip of paper. An 8.5 in. × 14 in. piece of paper can hold 4 or 5 such strips (along the 8.5 in. dimension). Alternatively, construct the strip using the words OBSERVATION → THEORY → (include the arrows) to emphasize the interaction of these two elements in the nature of science.

- *Optional Music:* "Circle of Life" from the *The Lion King* movie or other songs that feature the ideas of cooperation and community.

Points to Ponder

Teaching may be compared to selling commodities. No one can sell unless someone buys. We should ridicule a merchant who said he had sold a great many goods although no one had bought any. But perhaps there are teachers who think they have done a good day's teaching irrespective of what pupils have learned.

—John Dewey, American philosopher-educator (1859–1952) in *How We Think*

The strangest and best thing about teaching is that a seed dropped into what looks like rocky ground will often stick and take root gradually, and spring up years later...still carrying the principle of life.... The best kind of teaching...stops being the mere transmission of information and becomes the joint enterprise of friendly human beings who like using their brains.

—Gilbert Highet, writer and academic (1906–1978) in *The Art of Teaching*

Procedure

1. Give one of the double-sided strips to each learner. Point out to older students that sometimes teachers and learners are not on the same side of the page when it comes to explaining what their complementary roles and responsibilities are. Challenge the learners to individually manipulate the paper strip so that "TEACHING" connects to (or leads to) "LEARNING" (and vice versa). Allow a minute or two for learners to work on their own. Most learners will discover the solution by twisting the arrow head 180 degrees (or a half-twist) on one side of one end of the strip and connecting it with the other end. A piece of transparent tape can be used to secure the resulting Möbius strip so that it no longer has two distinct sides, but only one continuous surface. Have learners pull the taped strip between their thumbs and index fingers to see that this is the case. They could also hold a colored pen in the middle of the strip as the strip is pulled beneath it on a tabletop. Ask the learners for other tests they would like to perform on the Möbius strip; ask them also for their ideas for practical applications (some are cited in the Extensions section).

   (See answers to questions in step #2 on p. 23.)

2. *Optional:* If time permits, ask learners to construct a simple, untwisted cylindrical strip and compare it with the Möbius strip. Ask them to predict-observe-explain: What will happen if a hole is poked in the center of each of the two kinds of strips and a scissor is used to cut a straight line through the middle of the paper all the way around the strip? Also, what will happen as additional half-twists are made in Möbius strips made with longer strips of paper? See the Answers to Questions in Procedure section (p.xxx) for a related joke and limerick.

Debriefing

(See answers to Debriefing questions on p. 23.)

When Working With Teachers

Discuss how the Möbius strip is a visual participatory analogy for value-added, interactive teaching that leads to enhanced learning for both students and the teacher. *Note:* Studies conducted by the Education Trust (*www2.edtrust.org*) and other groups have found that as much as 40% of student variations in test scores can be attributed to teacher quality factors (versus 20% to small classes or small schools, 24% to parental college education, and 26% to other student background characteristics (e.g., poverty, language, and family composition). Simply put, there's no substitute for a great teacher! If time permits, discuss the following open-ended questions to elaborate on the analogy:

1. What kind of integrated curriculum, instruction, and assessment "twists" or strategies help learners construct cognitive and emotional connections to their teachers' (and textbook-based) knowledge?

2. How can teachers become learners in order to become better teachers? The *Drawing Hands* (1948) lithograph by M.C. Escher (see Internet Connections) can be used as another visual representation of a synergistic, interconnected feedback cycle among curriculum, instruction, and assessment.

3. Use the Dewey and Highet analogy-based quotes to explore the question, How can learners and teachers be interactive, connected collaborators who stay on the same side of the page?

4. How can the themes of the National Science Education Standards (Systems, Order, and Organization; Evidence, Models, and Explanation; Form and Function; and Constancy and Change [NRC 1996, p. 104]) be explored in the context of this activity?

When Working With Students

Beyond its use as an open-ended inquiry activity with a discrepant, unexpected "twist," the Möbius strip can be used as an analogy to help students understand their active roles as collaborative partners in learning with teachers.

Ask students how they intend to actively stay connected with, and provide feedback to, you, their teacher. Also, at appropriate points in the science curriculum, teachers can use Möbius strips as physical, analogical models for difficult to visualize, interactive science dualities such as MATTER and ENERGY or PARTICLES (electrons and photons) and WAVES (probability waves and light). Alternatively, a

Möbius strip with the words OBSERVATION and THEORY can be used to emphasize the two-way interactions of these elements of the nature of science.

Extensions

1. *Möbius and Mathematics Matter.* Explore the biography of the German mathematician Augustus Ferdinand Möbius (1790–1868) and various practical applications of topology. Möbius began his university studies with the intent of becoming a lawyer, but he was captured by astronomy and mathematics instead. The history of mathematics and science are intertwined—much like the Möbius strip. Many of the great innovators made significant contributions to both fields.

   Real-world applications of large Möbius strips include the following:

   - As long-lasting conveyor belts (that distribute wear and tear by using the entire surface area of the belt)
   - As continuous-loop recording tapes (to double the playing time)
   - In the manufacture of impact-based, computer printer ribbons (as well as the manufacture of now obsolete typewriter ribbons, which were twice as wide as the print head but allowed both half-edges to be evenly used)

   Möbius strips have also appeared in various artifacts of popular culture such as the following:

   - Some of M. C. Escher lithographs (e.g., *Möbius Strip II.* In that drawing—despite the fact that an endless ring-shaped strip usually has two distinct surfaces [one inside and one outside]—nine red ants crawl after each other and travel the front side as well as the reverse side. Therefore the strip has only one surface).
   - The international symbol for recycling, composed of three chasing arrows that form an unending loop.

- Martin Gardner's humorous short story "The No-Sided Pro- fessor" (in the book *Fantasia Mathematica* 1988)
- Arthur C. Clarke's science fiction story "The Wall of Dark- ness" (in *The Collected Stories of Arthur C. Clarke* 2002)
- Books such as Martin Gardner's *Mathematical Magic Show* (1978), which has a chapter devoted to Möbius strips, and Clifford Pickover's *The Möbius Strip: Dr. August Möbius's Mar- velous Band in Mathematics, Games, Literature, Art, Technology and Cosmology* (2006)

2. *Toys That Teach.* The idea of connected, interactive teaching and learning can also be visually modeled with the classic toy Er- nest the Balancing Bear, who pedals his unicycle back and forth from a position of higher gravitational energy to a lower height while balancing on a long monofilament line. Finding this cen- ter-of-gravity demonstration science toy may require a little web searching.

3. *Optical Illusions and "Reflecting" on the Science and Art of Teaching.* An *ambigram* (or *inversion*) is a graphical figure that spells out a word or phrase not only in its form as initially viewed, but also in another direction or orientation—although it may be either the same or a diametrically opposed message! As such, inver- sions are great tools for teaching the concept of symmetry. De- pending on the words or messages presented, inversions can also serve as visual participatory analogies that challenge teachers to consider alternative ways of looking at various aspects of the science teaching and learning dynamic. For example, reflect on the Teach-Learn and Upside-Down inversions of puzzle master Scott Kim (see Internet Connections); the paired complemen- tary words are designed to read the same when they are turned upside down as when they are right side up! Consider what cur- rent school practices might need to be turned "upside down" to increase the probability that our teaching efforts will result in enhanced student learning.

Note: Scott Kim's 125-page book, *Inversions: A Catalog of Calli- graphic Cartwheels* (1981, $14.95 ISBN: 978-1-55953-280-8), contains

more than 60 mind-boggling lettering designs based on various kinds of geometric symmetry; the book also includes essays about inversions to mathematics, art, psychology, and music and a section that shows students how to create their own inversions. See Activity #5 for a more extensive treatment of optical illusions.

Internet Connections

- Math Forum (how to make a Möbius strip): *http://mathforum.org/sum95/math_and/moebius/moebius.html*

- Mathematical Genealogy Project (link to a biography of August Ferdinand Möbius): *www.genealogy.math.ndsu.nodak.edu/html/id.phtml?id=35953*

- Official M.C. Escher Website (e.g., ties, posters, and T-shirts): *www.mcescher.com*

- Planet Perplex: *http://planetperplex.com/en/ambigrams.html* features a number of ambigrams including two of Scott Kim's copyrighted puzzles: Teach-Learn and Upside-Down. See also Scott Kim's Inversions—*www.scottkim.com/inversions/index.html*—for a gallery of his recent works (e.g., Input-Output, SuperTeacher, and Teach-Learn).

- Wikipedia: *http://en.wikipedia.org/wiki/Möbius_strip* and *http://en.wikipedia.org/wiki/Topology*

- Wolfram MathWorld: *http://mathworld.wolfram.com/Topology.html* and *http://mathworld.wolfram.com/MoebiusStrip.html*

Answers to Questions in Procedure, optional step #2

Cutting an untwisted cylindrical strip along the middle creates two separate cylindrical strips. Conversely, when a Möbius strip is cut, it forms a double-length continuous strip with two half twists. If the original Möbius strip is wide enough, you can even make one more cut on the double-length strip to produce two strips wound around each other!

When the number of half-twists is odd, the strip will be one-sided, whereas an even number of half-twists produce two-sided figures. The joke version of this puzzle is, Question: Why was Möbius such a strong debater? Answer: Because he truly believed there was only one side to every issue. Or, as a limerick: A mathematician confided/That a Möbius strip is one-sided/And you'll get quite a laugh/If you cut one in half/For it stays in one piece when divided!

Answers to Debriefing Questions

Analogies, brainstorming, concept mapping, discrepant-event demonstrations, experiments, and field trips are examples of minds-on instructional strategies that put a creative twist on conventional classrooms by keeping teaching connected to learning.

Teachers can also view their teaching analogically as "scientific research" in this way: "theory-informed hypotheses" guide their "experimental design" (i.e., integrated curriculum, instruction, assessment), "lab work" (i.e., teaching), and "data collection and analysis" (i.e., diagnostic, formative and summative assessments). The data collection and analysis in turn informs their subsequent research.

Activity 3

Burning a Candle at Both Ends: Classrooms as Complex Systems

Expected Outcome

A long candlewick is ignited at both ends and bent into a C-shape with the two burning ends placed 2–4 vertical inches apart. When the lower end of the wick is blown out, the flame from the upper end leaps downward to reignite it. (Instead of a candlewick, you can also use two candles. See Procedure, step #1.)

Science Concepts

For a fire to ignite, it is necessary to have fuel, an oxidizing agent (e.g., oxygen), and sufficient heat to reach the fuel's kindling temperature. A seemingly simple candle is actually a surprisingly complex, dynamic system that allows paraffin (a linear, alkane type of hydrocarbon) to burn after undergoing two phase changes. The room-temperature solid melts to form a liquid that rises up the wick via capillary action (from a pool that gathers at the base of the wick). The liquid on the wick then evaporates to form a combustible gas. Convection currents then cause the burning gas (and its carbon dioxide and water by-products) to rise while fresh, more oxygenated, cooler, denser air displaces the burning gas by-products mixture upward as they in turn feed the flame from below. When the bottom flame is extinguished, a trail of invisible, hot, unburned paraffin vapor (plus some visible, partially combusted carbon particles) continues to rise (for a brief time) and ignites if it encounters the top flame and then leaps downward to reignite the extinguished lower wick.

This long explanation is provided to emphasize that the age or developmental level of students (and the course content focus) should determine how deeply the teacher explores any scientific phenomenon. This demonstration can also be used to highlight the way that we can allow memorized answers to mask conceptual holes and misconceptions. For instance, the commonly used phrase "heat rises" causes many students to envision heat as a fluid (analogous to the incorrect caloric theory of the 1700s and early 1800s). It is actually more scientifically correct to say that, under the influence of gravity, cooler denser fluids fall and displace (upward) warmer, less dense fluids. But even this wording begs the question of how heat causes density differences. A given fluid is less dense when it is at a higher temperature because heat energy causes molecules to move faster and spread further apart in space (thereby decreasing the heated fluid's mass/volume ratio). Without this kinetic molecular theory perspective, convection may appear to happen magically by virtue of the assumption that "heat rises."

Science Education Concepts

This introductory activity models how simple it is to prepare and execute interactive, discrepant-event demonstration-experiments. They can be used on a daily basis to activate students' perceptual attention, catalyze cognitive processing, and energize interest in science phenomena. The predict-observe-explain (POE) instructional sequence elicits learners' prior knowledge, facilitates dialogue and active meaning-making, and develops science-mystery-solving skills that can uncover the scientific explanation behind various simple magic tricks. *Note:* American author Edgar Allen POE helped create the genre of mystery, science fiction, and horror short stories that included discrepant events and anomalies that are resolved by the end of the story, often with an unexpected twist—much like good science teaching but without the horror element!

This activity also serves as a *visual participatory analogy* (science education analogy) for classrooms as complex systems where teaches and students interact with one another and phenomena. Teachers can "light the fire" of interest in their sometimes under-motivated students; likewise, students who actively participate in learning can "relight" their occasionally burnt-out teachers. Similarly, teachers-helping-teachers forms of professional development enable peers to reignite each other's fire and passion for learning and for teaching.

Materials

- 1, 10–14 in. waxed wick (available from candle-making websites)
- OR 2, 6–12 in. candles
- OR make a wick by dipping a multi-strand cotton string into hot wax

Safety Notes

1. Make sure the work area is clear of combustible materials (e.g., paper) and flammable (e.g., oil, alcohol) materials.

2. Students and teacher should wear indirectly vented chemical splash goggles.

3. An ABC–type extinguisher should be available in case of emergency.

4. Caution students about working with active flames and high heat.

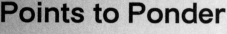

Points to Ponder

As a little wood can set light to a great tree, so young pupils sharpen the wits of great scholars: hence much wisdom have I learned from my teachers, more from my colleagues, but from my students most of all.

—Rabbi Hanina Ben Dosa (1st century AD) in the Talmud

The true aim of everyone who aspires to be a teacher should be, not to impart his own opinion, but to kindle minds.

—Frederick W. Robertson, English clergyman (1816–1853)

Education is not filling a pail but the lighting of a fire.

—William Butler Yeats, Irish writer (1865–1939)

Procedure

(See answers to questions in steps #3 and #5 on p. 33.)

1. Introduce the discrepant-event demonstration with a statement that includes a widely used verbal analogy: "People who are overworked [including teachers and students!] often feel that they are 'burning the candle at both ends.' Today we're going to study the science behind that phenomenon." For a more humorous effect, you might display this written message: "Today's lesson will focus on the physics and chemistry that occur when a slender illumination device is subjected to rapid carbonization on its antipodal points" and then ask, "What do you think we are going to be exploring?" In either case, light both ends of the long wick (or the exposed wick ends of two separate candles). For a more dramatic visual effect, turn the lights in the room off or down.

2. Bend the long wick into a C-shape (or place the flame of one candle just below the flame of a second candle) so that the two wick ends are 2–4 vertical inches apart.

3. Ask the learners to *predict* what will happen if the lower flame is blown out. What will happen if the upper flame is blown out? Do you expect the results to be about the same or different and why? What scientific principles support these predictions? Receive the learners' ideas without commentary.

4. Start by blowing out the upper flame. After nothing unexpected happens, reignite the extinguished upper wick by touching it to the lower flame. Then repeat the process, except blow out the lower flame. In this case, the learners will *observe* that the flame on the upper wick will leap downward to reignite the extinguished lower wick. (*Note:* Be sure to adjust the position of the upper burning wick to align with the rising vapors from the extinguished lower flame. This will only work if the airflow in the room is not too turbulent, so stand away from ventilation ducts and fans, which will disperse and dilute the rising unburned hydrocarbon vapors.)

5. Elicit a variety of initial, possible, plausible *explanations* of this phenomenon. Do not prematurely focus on the one "right" answer; let learners wrestle with the questions posed by the phenomenon. Explaining how a candle "works" (including this surprising effect) involves a number of science concepts that might cut across several instructional units.

Debriefing

When Working With Teachers

This demonstration can be used as a prop at the start of a professional development program or course to express empathy for the demands of teaching that cause teachers to sometimes experience burn-out. (If you use the alternative scientific-sounding description in Procedure, step #1, point out that facility with science terminology does not necessarily mean that someone understands the science behind a particular phenomenon.)

Use the quotes (all of which use the analogy of learning and fire) to discuss the value of daily use of interactive demonstrations and/ or experiments to reignite both students' and teachers' interest in and understanding of science. Also, discuss the analogical connec-

tion of both informal and formal teacher-to-teacher collaborations. If time permits, briefly discuss the not-so-obvious science behind a burning candle. You can also have teachers read the detailed explanation under Science Concepts and/or explore the resources cited in Internet Connections as their homework.

The use of discrepant-event demonstrations for teaching can be traced at least as far back in history to Michael Faraday's famous public Christmas Lectures for England's Royal Society in the mid-1880s. In fact, Faraday presented a series on the chemistry and physics of flames predicated on his belief that "There is not a law under which any part of this universe is governed which does not come into play and is touched upon in these phenomena. There is no better, there is no more open door by which you can enter into the study of natural philosophy than by considering the physical phenomena of a candle." His book *The Chemical History of a Candle* (1860), from which the quote in the previous sentence was taken, is still in print and offers many other intriguing activities (see Internet Connections). One is the use of a glass tube to channel off some of the unburned hydrocarbon vapors from the dark portion of a burning candle flame. These vapors can be ignited at the other end of the tube, thus getting two flames from one end of a single candle! Call it natural philosophy or science—this is still a magical experience!

When Working With Students

This demonstration could be used as either an Engage- or Elaborate-phase discrepant-event activity in 5E-based units that deal with the physics of convection and/or the chemistry of phase changes and combustion (i.e., oxidation of hydrocarbons). (For a discussion of the 5E Teaching Cycle, see Appendix B.) If used as an Engage-phase activity, don't use the Explain-phase teacher "answers" until after the mysterious phenomena has been further studied in the Explore phase (see, for example, Activity #14 in this book for a simpler, related convection activity that does not involve combustion). If the demonstration is used during the Elaboration phase, students should be able to explain the reigniting phenomenon as dependent on convection currents that push hot unburned paraffin gas upward. Ask students to predict whether convection experiments would work in a zero- or microgravity environment and how they might test that (see the NASA site listed in Internet Connections to study the effect of "turning off" gravity and convection currents during the free fall of a burning candle).

A variety of other real-world application questions can be used as follow-ups. For example, challenge students to consider why the direction of spin of room ceiling fans is different for summer and winter. (In summer the fan is set on "forward"; in winter the fan is set on "reverse" in order to push warm air down to the floor where people are.) Ask why heaters are better placed on the floor than the ceiling (and just the opposite for air conditioners).

Extensions

1. *Burning the Candle at Both Ends, Version #2* (see Internet Connections). The teacher exposes the wicks at both ends of a 10–12 in. long, 0.5–0.75 in. diameter candle and inserts a hot metal rod or knitting needle through its center of mass. Balance the candle on the metal rod across two large glasses so that it is free to rotate in the vertical direction. Light both ends of the candle and in a few minutes it will begin to oscillate about the pivot point with a regular frequency, making for a surprising and "edu-taining" science conversation starter. If desired, use a black marker to label one end "T" (for Teacher) and the other "S" (for Student) to emphasize the give and take of effective classroom interactions.

2. *"Magic" Birthday Candles.* These candles, which self reignite when blown out, are fun to use with students as a related discrepant event. (They also serve as a visual analogical model for the "in-extinguishable" energy of dedicated teachers....) (See Internet Connections: How Stuff Works.)

3. *The Edible Candle.* This demonstration (see Internet Connections: Terrific Science) uses a cylinder-shaped piece of apple, potato, or banana dipped in lemon juice (to prevent discoloration from oxidation) with a fresh, oil-rich almond-sliver "wick" to make a convincing, flammable, edible candle that burns for several minutes (until the almond's oil is consumed). Alternatively, a cylindrical piece of string cheese can be used as the candle body.

    This entertaining discrepant event can be used to emphasize the difference between observations and inferences. It is especially

relevant in a biology unit on cellular respiration to dramatically (and humorously) emphasize Antoine Lavoisier's (French chemist, 1743–1794) discovery that respiratory gas exchange involves a combustion or oxidation process that is analogous to a candle burning. Of course, humans do not have the ability to burn hydrocarbons like paraffin, but we do "burn" carbohydrates, fats, and proteins.

4. *Homemade Hot Air Balloons and Lava Lamps.* See the two Extensions for Activity #14, which also focus on density differences and convection.

5. *Science Mentor.* Michael Faraday's mentor, Sir Humphrey Davy, also studied the physics and chemistry of flames and invented (and deliberately did not patent) the coal miner's safety lamp (see Internet Connections: Whelmers #42).

6. *Misconceptions Matter.* Teachers can use any of the following references to read about the nature of student misconceptions and their implications for integrated curriculum, instruction, and assessment: Driver, Guesne, and Tiberghein 1985; Driver et al. 1994 (especially useful); Duit 2009; Fensham, Gunstone, and White 1994; Harvard-Smithsonian Center for Astrophysics (i.e., MOSART); Keeley, Eberle, and Farrin 2005; Keeley, Eberle, and Tugel 2007; Kind 2004; Meaningful Learning Research Group; Olenick 2008; Operation Physics; Osborne and Freyberg 1985; Science Hobbyist; Treagust, Duit, and Fraser 1996; and White and Gunstone 1992.

Some authors use related terms that are somewhat less judgmental than *misconceptions* (e.g., *preconceptions, alternative conceptions, naive conceptions,* or *children's science*) to refer to the ideas that students have about a concept before a teacher formally teaches it. These student-generated ideas usually reflect a creative (though typically *not* completely correct) blending of "lessons learned" from informal life experiences and from formal schooling. Students' misconceptions are commonly activated when discrepant-event activities are used; thus they are relevant in all subsequent activities in this book. Several of this book's activities are specifically designed to focus teachers' attention on this important topic (i.e., activities #20, #24, #27, #28, and #29).

Internet Connections

- Burning the Candle at Both Ends, Version #2. From Martin Gardner's 51-page Science puzzles—Gravity machine activity: *www.vidyaonline.net/arvindgupta/martingardner.pdf*

- HowStuffWorks: Candle: *http://home.howstuffworks.com/question267.htm* and Trick birthday candles: *http://science.howstuffworks.com/question420.htm*

- NASA: Candle flame in microgravity (classroom demonstration/experiment): *http://quest.nasa.gov/space/teachers/microgravity/9flame.html*

- Project Gutenberg: The chemical history of a candle: *www.gutenberg.org/etext/14474*. See also: *www.fordham.edu/halsall/mod/1860Faraday-candle.html*

- Terrific Science: Edible candle: *www.terrificscience.org/movies/pdf/illuminating_demo.pdf*

- University of Iowa Physics and Astronomy Lecture Demonstrations (video clips; five demonstrations on convection under Heat and Fluids): *http://faraday.physics.uiowa.edu*

- Whelmers #42: Fire sandwich (a model of Davy's coal miner's safety lamp): *www.mcrel.org/whelmers/whelm42.asp*

Answers to Questions in Procedure, steps #3 and #5

The detail and depth of explanation should, of course, be appropriate for your students and their prior understandings. A short version of the explanation under Science Concepts (on p. 26) is that paraffin actually burns in the gaseous state, and it is the trail of the partially unburned, hot paraffin vapors rising via convection currents that re-ignites when the lower candle (or wick end) is extinguished.

Section 2:
Human Perception as a Window to Conceptions

Activity 4

Perceptual Paradoxes: Multisensory Science and Measurement

HOT WATER

COLD WATER

ROOM TEMPERATURE

A

B

C

Expected Outcome

Three experiments are performed with volunteers that both challenge assumptions about the human sensory system's fidelity and point to the need to use multiple senses and quantitative measurements to get a more accurate representation of reality.

Science Concepts

Human sensory perceptual system—seeing, hearing, tasting, smelling, and heat flow and pressure-sensitive touching—function synergistically to help us form generally reliable, adaptive, internal concepts of our external environment. However, *misperceptions* are one of several sources of *misconceptions* that can be quite tenacious and resistant to change. Quantitative measurements taken with scientific instruments can help overcome human sensory biases, subjectivity, and limited sensory ranges. Scientists need to trust, test, and transcend human sensory perceptions.

Science Education Concepts

Multisensory, minds-on activities (including hands-on explorations) provide the experiential foundation for new learning and, when necessary, for reconstruction of prior understandings. However, the truism "seeing is believing" suggests an unproblematic, one-way flow from perception to conception. Perceptual paradox activities challenge this assumption in ways that are both playful and mentally challenging and that point to the need for quantitative measurements. Perceptual paradox activities also demonstrate that human meaning-making always involves both selective perception and active conception-making. The latter is a cognitive act that can, at times, be misled into drawing erroneous conclusions for a variety of reasons, including perceptual biases and limitations. Human perceptions can be viewed metaphorically as a somewhat foggy window between the outer, external world and the inner world of our individually constructed conceptions (analogous to Plato's famous "Allegory of the Cave" in *The Republic*). More specific *visual participatory analogies* are developed for each of three experiments in the Debriefing section.

Safety Notes

1. Students and teacher should wear indirectly vented chemical splash goggles during this activity.
2. Teacher and student volunteer should wear aprons.

Materials

Experiment #1
- 3 identical-size containers large enough to submerge a hand in water (e.g., rectangular, 12 in. × 7 in. × 1.5 in. casserole dishes work nicely)
- Thermometer (non-mercury)

Experiment #2

- Bite-size slices or cubes of both apples and potatoes for "blind" taste testing

Experiment #3

- 2 empty coffee cans: one a 1 lb. can and the other a 3 lb. can. Fill the 1 lb. can with sand. Put slightly less sand in the 3 lb. can so that the total weight of the 3 lb. can is the same as the filled 1 lb. can. Cover with the opaque plastic lids from the 2 coffee cans.
- Scale or balance

Points to Ponder

Great is the power of misrepresentation: the history of science shows that fortunately this power does not long endure.

—Charles Darwin, English naturalist (1809–1822) in *On the Origin of Species* (1859)

The brightest flashes in the world of thought are incomplete until they have been proved to have their counter-parts in the world of fact.

—John Tyndall, Irish physicist (1820–1893) in *Fragments of Science for Unscientific People* (1871)

Procedure

Introduce the following activities by mentioning that evolution by natural selection shaped human sensory systems that work on a "need-to-notice" basis to help us access and process environmental data related to our survival and reproduction as a species. The systems also form the foundation of scientific knowledge. Raise these focus questions with your students: Do our senses ever "lie" to us or misrepresent reality? Is it true that "to err is human"? (See answers to the questions in experiments #1–#3 on pp. 45–46.)

Experiment #1: Hot or Cold or Just About Right?
Thermal Energy and the Goldilocks Effect

1. Place equal volumes of ice-cold (0–5°C, but no ice), room temperature (~22°C), and hot tap water (~50°C) into three identical containers, placed in order from left to right next to each other. Do not indicate to the learners that the three containers are at different temperatures. (*Safety Notes:* Make sure the hot water is not so hot that it could burn or otherwise injure the volunteer. The teacher may wish to use a glove when handling the container with hot water.)

2. Ask a volunteer to simultaneously place his or her left hand in the ice-cold water and his or her right hand in the hot water, simultaneously, for 1–2 min.

3. Then ask the volunteer to place both hands in the middle container of room temperature water for another 1–2 min., remove his or her hands, and mentally note the sensory experience. Repeat steps #2 and #3 with at least one other volunteer. Ask the volunteers to orally describe their experiences to their peers.

4. Ask: Is the human skin an accurate instrument for measuring temperature? If not, what does the human skin actually measure? Use a thermometer to measure the temperature of water in the three containers. Ask: Are heat and temperature the same thing or different? If they are different, how do scientists distinguish and measure the two concepts? Could a body of water be at a higher temperature, but contain less heat, than a second body of water at a lower temperature?

Note: An alternative or supplemental experiment is to ask all the learners to place their right hands first on a plastic or wood portion of their desks or chairs and then on a metal portion. Although the entire desk is at the same temperature as the room, the metal "feels colder" than the other parts because it is a better conductor of heat than wood, plastic, or human skin. When learners touch the metal, body heat flows into the metal from the skin, which then feels cool.

Experiment #2: A Taste-Less Trick?

1. Prepare identical shape and size slices (or cubes) of apple and potato with the skins (and the core of the apple) removed. (If desired, onions could also be used.) Do NOT cue the class into what food items you have prepared, but ask for several volunteers to participate in a "blind and smell-free" taste test. Ask the volunteers to close their eyes (or use blindfolds) and hold their noses while they sample and try to identify what they are eating. Ask: What was/is the evolutionary importance of being able to accurately identify possible foods by sight, smell, and/or taste?

2. Repeat the experiment, but allow the volunteers to breathe (and smell) normally as they eat.

3. Allow them to look at the foods and compare the accuracy of the predictions. How might we account for identification failures? Are two (or three) human senses sometimes better than one? What scientific instruments do chemists use to identify the individual components of complex mixtures such as foods, perfumes, and coloring agents?

Experiment #3: Can You Decide a Weighty Matter?

1. Tell the class that you are testing their sensitivity thresholds for detecting weight differences (this is somewhat of a misdirection).

2. Ask a series of volunteers to come to the front of the room and to pick up the smaller can of sand and then the larger can of sand. Ask them to individually write down which can they think is heavier.

3. Poll the volunteers and tally the results. Ask: What scientific instrument would we use to answer the question of which is "heavier"? Proceed to use a spring scale for weight (or a balance for mass) and share and discuss the discrepant results. What else might we be sensing when we use our hands to weigh the two cans?

> ### Safety Note
>
> Have student volunteers wash their hands with soap and water prior to touching food. Make sure that food is placed only on hygienically clean surfaces that are free of chemical residue and dirt.

Debriefing

The two main messages to convey to both teachers and students are as follows:

1. All acts of cognition involve selective perception, encoding, and reconstruction of sensory data into conceptual frameworks. The frameworks are influenced by the learner's unique prior knowledge and abilities, which determine the mix of assimilation versus accommodation that is necessary for learning.

2. Some limitations and biases of our sensory systems can be overcome by using scientific instruments to make quantitative measurements.

When Working With Teachers

Point out that these perceptual paradoxes provide counterexamples that expose the erroneous pedagogical assumption that learning consists of passive recording of external stimuli. Each of the experiments above serves as a unique visual participatory analogy for science education that can be discussed and critiqued.

Experiment #1

How different students respond to what appears to be the same educational environment can vary based on their unique multiple intelligences (MI) profiles and the lessons they have learned in prior educational environments. Analogically, a challenge that is "just the right temperature" for one student—that is, in his or her zone of proximal development (ZPD)*—may be "too hot" (intellectually demanding) for one student or "too cold" (boring) for another. Differentiated curriculum, instruction, and assessment connect to students' diverse MI profiles by using tasks with appropriate cognitive "loads" that support learners as they move from where they are to the desired learning outcomes. Teachers who wish to explore these topics in more depth can look to the Internet Connections.

---

*A learner's ZPD is what the learner can achieve—based on prior knowledge and abilities—when scaffolding is provided by a carefully targeted instructional sequence and a supportive teacher.

Experiment #2

This experiment can be used as analogy for how over-reliance on text-book terminology, or a facts-first-phenomena-follow approach, can result in bland science teaching and overfed but undernourished students.

Experiment #3

This experiment can be used as an analogy for cognitive load or "conceptual density" (i.e., the number and "weight" of distinct concepts that are covered in a given amount of instructional time) relative to the learner's readiness or ZPD. If more time is allocated for a given number of "big ideas" (i.e., analogous to the weight of the sand being spread out over a larger surface area), students will perceive that their learning is a lighter, more sensible load. Both the National Science Education Standards (NRC 1996) and the Benchmarks for Science Literacy (AAAS 1993) argue for a "less is more" approach that challenges students to use inquiry to uncover nature's secrets rather than to simply memorize a list of scientific terms.

Extensions

1. Explore various types of optical illusions and websites in Activity #5 in this book.

2. *Silly Cylinders.* Science supply companies sell equal mass, unequal volume cylinders (i.e., different density metals and plastics) as well as unequal mass, equal volume cylinders that will surprise students when they compare their handheld impressions to the results from a balance.

3. *A Penny for Your Thoughts.* An inexpensive challenge activity is to see if students can separate pre-1982 pennies (nearly pure copper) and post-1982 pennies (zinc inside with a thin copper coating) by hand when blindfolded (the difference is 0.5 g). A mixture of these two types of pennies can also serve as an analogy for different mass isotopes of a given chemical element.

Internet Connections

(Also see the Internet Connections in Activity #5, pp. 59–60.)

• Athabasca University's Centre for Psychology Resources

(AUPR): Smell and taste links: *http://psych.athabascau.ca/html/aupr/sensation.shtml#Smell%20and%20Taste*

- CAST: Differentiated instruction: *www.cast.org/publications/ncac/ncac_diffinstruc.html*

- Exploratorium Snacks: Your sense of taste (as related to smell; uses candy): *www.exploratorium.edu/snacks/your_sense_of_taste/index.html*

- Human Intelligence: New and emerging theories of intelligence (e.g., synopses of Gardner's theory of multiple intelligence, Sternberg's theories, Goleman's concept of emotional intelligence): *www.indiana.edu/~intell/emerging.shtml#intro*

- International Mind Brain and Education Society: *www.imbes.org*

- Reframing the Mind: A critique of the theory of multiple intelligences (Daniel Willingham): *http://educationnext.org/reframing-the-mind*. See also other articles and critiques of brain-based education fads by the same author: *www.danielwillingham.com*

- Society for Neuroscience: Brain facts: A primer on the brain and nervous system (74-page booklet, CD, and free pdf file): *www.sfn.org/index.aspx?pagename=brainfacts*

- Howard Hughes Medical Institute (HHMI): Seeing, hearing, and smelling: Illusions reveal the brain's assumptions: *www.hhmi.org:80/senses/a110.html*

- HowStuffWorks: Human senses: *http://health.howstuffworks.com/question242.htm* and Taste impairment: *http://healthguide.howstuffworks.com/taste-impaired-dictionary.htm*

- The History Guide: Lectures on modern European intellectual history: Plato, The Allegory of the Cave, Book VII of The Republic: *www.historyguide.org/intellect/allegory.html*

- University of Virginia Physics Department: Mixing warm and cold water lab: *http://galileo.phys.virginia.edu/outreach/8thGradeSOL/WarmColdFrm.htm*

- Wikipedia: Cognitive load theory: *http://en.wikipedia.org/wiki/Cognitive_load_theory* and Zone of proximal development: *http://en.wikipedia.org/wiki/Zone_of_Proximal_Development*

Answers to Questions in Procedure, Experiments #1, #2, and #3

Experiment #1. The volunteers' brains will receive two conflicting messages: Their left hands will describe the room temperature water as warm and their right hands will describe it as cold. Unlike a thermometer, human skin responds to heat flow. *Hot* and *cold* are relative terms; we experience something as hot if it transfers heat to us and we experience something as cold if we transfer heat to it. This transfer is dependent on the relative temperature and conductivity of the substance compared to our hands. If the substance is relatively warmer, heat will flow to our skin. If the substance is a better conductor, heat will flow out from our skin. Heat is a measure of total kinetic molecular energy as measured in calories or joules with a calorimeter. Temperature is a measure of average kinetic molecular energy as measured in degrees Celsius, Fahrenheit, or Kelvin with a thermometer or infrared sensor. Given the differences between the two concepts, a cup of boiling hot water can have a higher temperature, but lower heat content, than a much larger volume of lukewarm water.

Experiment #2. In the evolutionary history of the human race, the fact that people had a refined sense of smell was adaptive in helping them to avoid spoiled food or toxins that could harm them before they actually took the risk of putting the food into their mouths. Good olfactory intelligence also helped people to identify and select high-energy, high sugar, and/or vitamin rich foods without any conscious knowledge of nutritional biochemistry. The nervous systems of all organisms are geared to "notice on a need-to-know" basis. Most other mammals (e.g., dogs) have a much more refined sense of smell (and hearing) than *Homo sapiens*. Our 5,000–10,000 taste buds distinguish between sweet, salty, sour, and bitter. These sensations relate to chemically measurable quantities of hydroxyl groups (OH) in sugars, metal ions in salts (e.g., Na^+ in table salt), acids (H^+) in sour-tasting foods, and alkaloids ($^-$OH) that trigger the bitter receptors. Much of the sensory experience of "taste" is actually an interaction of taste, smell, and tex-

ture (i.e., sensory interdependence). This is why most people lose the ability to "taste" food when they have a cold or sinus infection.

Experiment #3. Humans can confuse perceived weight with density. In this case, most volunteers will experience the smaller can as being heavier (because it is denser) than the larger can of equal weight. This misperception-based misconception is likely due to the pressure-sensitive nature of touch (i.e., the smaller can exerts a greater weight per unit area of our hand). In this case, "believing is feeling."

Activity 5

Optical Illusions: Seeing and Cognitive Construction

image by travelpixpro for iStockphoto

Expected Outcome

Various optical illusions are displayed and different learners report that they "see" very different things in the same image; can't make any sense at all of the image at first examination; or see different things depending on what other visual elements are immediately adjacent or whether they make quantitative measurements.

Science Concepts

"Seeing" involves both the camera-like, mechanical role of the eye (whose lens focuses incoming light and projects an upside down image on the retina) and the pattern finding, sense-making capability of the brain. The brain cognitively "focuses" on certain holistic combinations of elements in an image and interprets and constructs "what it all means" based on prior experiences and current context. With nearly half of the brain's cortex devoted to processing visual stimuli, these dual processes of perceiving and conceiving normally occur unconsciously, seemingly effortlessly, very quickly and highly reliably to construct meaning based on information about shape, color, spatial orientation, location, and movement. Visual intelligence often precedes and always interacts with our rational, logical-mathematical ("head") and emotional or intra- and interpersonal intelligences ("heart").

Scientists depend heavily on visual information and use technology to extend the range of our sight to include the worlds of the very small, large, slow, fast, and even different portions of the electromagnetic spectrum with frequencies much higher and lower than visible light that can be detected by the human eye. Given that we are a highly vision-dependent species, marketing and advertisements rely heavily on visual images to capture consumers' attention and influence their subsequent emotional and behavioral responses (e.g., "seeing is believing... a picture is worth a thousand words... can you see yourself owning...").

Four Types of Optical Illusions

1. *Ambiguous figures* are intentionally designed to confound our visual intelligence in that they can be seen as having multiple, quite different images in one drawing. Two of the most famous of these illusions were both made popular in the year 1915. When cartoonist W. E. Hill's "my wife and mother-in-law" image is looked at one way, you see the face of a beautiful young woman turned slightly askance, but looked at from another perspective, you see the face of an old woman with a big nose (Figure 1). The Planet Perplex website credits the original source to ads used during the late 1800s (see *http://planetperplex.com/en/ambiguous_people.html*).

Another famous figure-ground type of ambiguous figure is the (Edgar) Rubin vase or goblet/two faces image (see Figure 2). This illusion has been traced back to a 1795 picture puzzle entitled *Weeping Willow* (Hoffman 1998, cited in the Materials section). Variations on these types of figures are images that appear as something completely differently when rotated 180 degrees or turned upside down.

Figure 1

Source: *www.qualitytrading.com/ illusions/girlwoman.html*

Figure 2

Source: *http://en.wikipedia.org/wiki/ Rubin_vase*

2. *Anamorphoses proper* were developed nearly five hundred years ago by the artist Erhard Schon and others. In this technique, distorted sketches of famous people were drawn vertically elongated so when looked at from a downward angle, they could be seen in the proper perspective and recognized. Catoptric images are a related art form. In order to be seen, a catroptric image must be reflected in a distorting "magic" mirror (e.g., cylindrical, conical, and pyramidal) (see *www.anamorphosis.com/index. html*). Computer software such as Microsoft's WordArt can easily create tilted image messages with words that can only be read if held at an angle.

3. *Perceptual set optical illusions* demonstrate the importance of both context and prior experiences in human cognition. The information gaps in images made up of dots; incomplete, broken lines; or

inkblots are filled in by viewers who activate appropriate memories to help them make sense of the partial images. Viewers unable to do this without assistance often can be given a verbal cue that will enable them to focus in on the answer. Typically, these illusions are designed to elicit only one "correct" image, but Rorschach inkblot images (as used in psychoanalysis) are deliberately open to a wide range of interpretations. The latter are associated with the phenomenon of *pareidolia* where people see or read patterns into images (and/or random sounds) where no pattern was intended (e.g., this often takes the form of people seeing religion-linked images in everyday objects such as clouds, rocks, astronomical photographs, and even baked goods). See the Internet Connections for examples and more details on these "believing is seeing"–type phenomena. The role of context is also demonstrated by cases in which the same letters and numbers are read differently based on adjacent visual elements that elicit one kind of interpretation versus another.

4. *Shape and/or linear size distortion illusions* involve instances where the eye is tricked into believing that different objects or lines are the same or different sizes (than they really are) or are parallel (or not). These are instances where objective quantitative measurements reveal that the unaided, naked eye has been misled by perspective or context.

Science Education Concepts

Optical illusions can be used as interactive, everyone-participates demonstrations to show teachers that the act of visual observation (much less the conceptual learning we expect in our science classrooms) is not a passive, mindless, stimulus-response, camera-like operation. Seeing involves active, minds-on processes of stimuli filtering, selecting, encoding, and meaning-making (i.e., cognitive construction and interpretation). Normally, we do this so seemingly effortlessly that we do not consciously notice the process (and, as a result, are not aware of our sensory biases). But optical illusions challenge our simple notions of "seeing is believing" and suggest that at times, it is more appropriate to say that "believing is seeing." Research on optical illusions was begun by German "Gestalt" psychologists during the period between the two world wars (see Internet Connections: Gestalt Principles). Gestalt psychology is most commonly associated with the concept of emergent properties (i.e., the whole is greater than the sum of the parts).

Optical illusions can also serve as *visual participatory analogies* for the minds-on, constructive, emergent nature of all learning. The brain constructs meaningful conceptions from sensory perceptions by forming educated, value-added best guesses about the nature of reality as informed by prior knowledge and selective sensory inputs that filter and color the reality we experience. This process involves both inside→out or top→down (brain→world) processes and the reverse, outside→in or bottom→up (world→brain) processes. The brain does not function like a video recorder but instead is constantly "rewiring" neural connections and reconstructing our ever-changing understandings and how we perceive past, present, and future events and phenomena. Developing skills of careful observation (e.g., quantitative measurements) and unbiased interpretation is central to science and minds-on thinking that is associated with learning that lasts. Regular use of visually stimulating "fun" phenomena is an essential element of high-quality science teaching.

Materials

Obtain samples of the four types of optical illusions from books such as the following:

- Block, R., and H. Yucker. 1989. *Can you believe your eyes? Over 250 illusions and other visual oddities*. New York: Routledge. (This book includes explanations of the science underlying the illusions.)
- Hoffman, D. 1998. *Visual intelligence: How we create what we see*. New York: W.W. Norton. (This book includes explanations of the science behind visual intelligence.)
- Kay, K. 1997. *The little giant book of optical illusions*. New York: Sterling. (311 illusions)
- Seckle, A. 2005. (reprint). *The great book of optical illusions*. Willowdale, ON, Canada: Firefly Books.
- Seckle, A. 2009. *Optical illusions: The science of visual perception*. Willowdale, ON, Canada: Firefly Books. (275 illusions)

Alternatively, download images (see Internet Connections websites) and make your own collection of images to project (or distribute to students in the case of illusions that require linear measurements). Two especially useful sites and some of the sample illusions they include are the following:

1. *www.optillusions.com* Samples include American Indian face/ back view of the whole body of a hooded Eskimo (see Figure 3, p. 54); duck/bunny; woman's face/male sax player; mother/ father/daughter wearing a common beret (see Figure 5, p. 54); and a negative image of a man/the word *Liar* in script writing.

2. *http://planetperplex.com/en/index.html* This site has some of the same images as those listed above, plus many others.

(*Note:* Some of the websites contain interactive illusions and vibrant full color images—much more interesting than the static, black-and-white images in books.)

Points to Ponder

Most of the knowledge and much of the genius of the research worker lies behind his selection of what is worth observing. It is a crucial choice, often determining the success or failure of months of work, often differentiating the brilliant discoverer from the... plodder.

—Alan Gregg, English scientist and author (1941–)

What we want to see is the child in pursuit of knowledge and not knowledge in pursuit of the child.

—George Bernard Shaw, Irish playwright (1856–1950)

Procedure

Exercise #1: Ambiguous Figures

1. Project in sequence several of the classic ambiguous figure illusions and ask the learners to silently keep an ordered, written record of what they see. After projecting the last one, review and tally the class results.

2. (For answers to the questions in this step, see p. 61.) As you begin to get discrepant results (i.e., some learners see one thing, others another), ask the following questions: Does this experience support the theory that the act of seeing is a passive process analogous to a camera or camcorder taking a picture? If not, could seeing involve some form of selective attention and filtering that might amplify, mute, and differentially combine the same external visual data? Does seeing involve some form of very fast and unconscious cognitive construction of meaning via pattern recognition? Do our prior experiences and conceptions influence what we see? (*Note:* The same visual data or stimuli are being projected to everyone, yet they elicit different interpretative responses.)

3. Focus attention on one or two of the images where an alternate image is seen by only a minority of the learners (Figures 3, 4, 5). Use a pointer to draw attention to certain visual elements (or use an opaque cover to block out specific portions of the projected image). With practice and guided directions, most people should be able to switch their perception to enable them to see the alternate image. Yet, no one will be able to see both interpretations at once!

Figure 3

The head (only) of an American Indian looking to the left and the whole body of an Eskimo looking to the right (facing away). This is also called the Wilson figure.

Source: *http://planetperplex.com/ en/ambiguous_people.html*

Figure 4

Appears as a swan facing left or as a squirrel facing to the right. (G. H. Fischer 1968)

Source: *http://planetperplex.com/ en/ambiguous_animals.html*

Figure 5

This image shows a father, a mother, and a daughter who all share the same beret-style hat. The father is looking to the front right and has a big nose and a moustache. The mother is looking to the front left and is wearing the "moustache" as a scarf. Her left eye is the father's right eye. The daughter is looking to the back left. The eye of the father is her ear. (G. H. Fischer 1968)

Source: *http://planetperplex.com/en/gh_fisher.html*

Exercise #2: Anamorphoses Proper (or Tilted Image Drawings)

1. Make your own images with Microsoft Word by going to Word-Art and picking a solid color, linear font. Type your message of one or more words and then drag the selection bar to compress the image in the horizontal (and elongate it in the vertical) direction. Make an overhead transparency or PowerPoint slide for projection and photocopies for distribution.

2. Project a tilted image sentence (e.g., "Science is FUNdaMENTAL" or "Look at science from a different angle") and tell the learners that each contains a hidden verbal message with a specific number of words (tell them that number). Viewed from this flat, vertical perspective, no one is likely to be able to see and read the message. Distribute photocopies of the same image and suggest that sometimes when we can manipulate an object, we can see it differently (*Teacher Note:* This is one argument for hands-on explorations). After a minute or two, if the "trick" hasn't been discovered, suggest that in science sometimes we need to look at things from a different "angle" or perspective to discover previously unnoticed patterns. If you need to, demonstrate how to hold the photocopy at an angle slanted down and away from your eyes. Also, if you have the means to project a WordArt file from a computer, you can morph the hidden message to varying extents live before their eyes, making it harder or easier to read on a flat vertical screen.

Exercise #3: Perceptual Set Optical Illusions

1. Project partial images (alternate showing the vertical and horizontal sequence) of optical illusions, before showing the entire images such as in Figure 6, page 56.

Figure 6

The horizontal sequence reads as the letters ABC, while the vertical sequence reads as the numbers 12, 13, and 14. What we see in the middle depends on the context of what surrounds it. This image appears in many books and websites including *http://planetperplex.com/en/item97* and *www. qualitytrading.com/illusions/letters.html.*

Exercise #4: Shape and/or Linear Size Distortion: Measurements Matter

A fun and mentally engaging way to develop skills of linear metric measurements in a real-world, problem-solving context is to provide learners with copies of a variety of optical illusions that involve size comparisons. Ask them to compare and contrast unaided, naked-eye observations to quantitative measurements. Working to resolve such perceptual puzzles is far more engaging than the standard approaches to teaching linear measurement skills without a context or puzzling problem to solve. Furthermore, it simultaneously impresses on learners the need for empirical evidence, logical argument, and skeptical review. Quantitative measurements add a level of objectivity to observations that can otherwise be quite subjective (and mistaken). In Figure 7, the height of the top hat is actually the same as the width of the hat, even though our eyes suggest that the hat is taller than it is wide (Fick 1851).

Figure 7

Source: *http://planetperplex.com/en/a_fick.html*

Debriefing

Both teachers and students benefit from understanding how visual intelligence functions as an example of the active, minds-on nature of learning as a constructive act of meaning-making. The two quotes on page 53 point to the relevance of this fact to both scientific research and teaching. Making sense of new learning experiences is influenced by our prior experiences and conceptions and the particular contexts in which the new experiences are encountered. The science historian Thomas Kuhn suggested (in his 1962 book, *The Structure of Scientific Revolutions*) that "normal" science evolves gradually by slow accumulation of observations and refinement of theories guided by a dominant paradigm or overarching perspective.

However, eventually anomalies arise that can only be resolved with a revolutionary new paradigm that causes previous ideas to be substantially reframed. In this respect, the history of science parallels the conceptual changes that occur in learning. Much of what we consider learning involves the assimilation of new information into our preexisting mental schemata in a manner analogous to filling in missing pieces of a puzzle. However, paradigm shifts also occur in the minds of individual learners when they are challenged to give up previously held misperceptions and/or misconceptions. Accommodation of new experiences that do not fit into preexisting paradigms requires considerable mental effort because we need to unlearn and re-conceptualize what we thought we understood (i.e., we have to reorder our partially completed puzzle to form a different picture).

Teachers and students have complementary responsibilities to activate and reconsider prior conceptions and to look for new conceptual connections. Learning science is greatly facilitated by teachers systematically using minds-on, discrepant-event demonstrations; hands-on explorations; and analogies to try to establish both a need and common experiential base for (re)constructing conceptual networks of understanding ("phenomena first, facts follow"). Critical analysis and group discussion of such experiences provide essential diagnostic and/or formative assessments to determine if the different learners made similar observations and formed related inferences (or, if not, determine how to modify subsequent instruction to help guide everyone to "get it").

Extensions

1. *Optical Illusions, Neurobiology, and Cognitive Science Research.* The books listed in the Materials section and Internet Connections provide many optical illusions and theoretical background for interested teachers and students. See also Activity #4 in this book for paradoxes associated with other senses, as well as Activity #28.

2. *Optical Illusions as Science Analogies.* The ambiguous figure illusions can be used as visual analogies for the concepts of wave-particle duality and matter-energy equivalence—situations in which scientists "see" one thing or another depending on the context but never both forms simultaneously.

Internet Connections

Internet websites contain a wealth of static and interactive illusions. Optical illusions are the most widely studied perceptual illusions, but sites such as the Athabasca University, Howard Hughes Medical Institute, and University of Colorado at Boulder home pages include illusions related to other senses as well.

- Al Seckel's Homepage of Illusion, Perception and Cognitive Science: *www.illusionworks.com*

- Archimedes' Laboratory: Illusions and puzzles: *www.archimedes-lab.org/sitemap2.html*

- Athabasca University's Centre for Psychology Resources (AUPR): Sensation and perception: *http://psych.athabascau.ca/html/aupr/sensation.shtml*

- Exploratorium Online Exhibits: 13 optical + 3 auditory interactive illusions: *www.exploratorium.edu/exhibits/nf_exhibits.html*

- Exploratory Science Centre: *www.exploratory.org.uk/illusions/flash/index.htm*

- Eyetricks.Com: *www.eyetricks.com*

- Gestalt Principles (tutorial related to graphic design): *http://graphicdesign.spokanefalls.edu/tutorials/process/gestaltprinciples/gestaltprinc.html*

- Howard Hughes Medical Institute (HHMI): Seeing, hearing, and smelling: Illusions reveal the brain's assumptions: *www.hhmi. org:80/senses/a110.html*

- Illusions-Optical.com (gallery of static and animated illusions): *www.illusion-optical.com*

- Illusion Works (includes Java applets and Shockwave interactive demonstrations): *www.psychologie.tu-dresden.de/i1/kaw/ diverses%20Material/www.illusionworks.com*

- Neuroscience for Kids: Gallery of visual illusions (interactive): *http://faculty.washington.edu/chudler/flash/nill.html*

- NIEHS Kids' Pages: Optical illusions: *http://kids.niehs.nih.gov/ illusion/illusions.htm*

- Official M.C. Escher Website: *www.mcescher.com*

- Optical Illusions: *www.optillusions.com*

- Optical Illusions: Collection of visual, scary, and funny illusions: *www.newopticalillusions.com*

- Planet Perplex: *http://planetperplex.com/en/index.html*

- Quality Trading.Com (39 public domain optical illusions): *www. qualitytrading.com/illusions*

- Sandlot Science: Optical illusions and brain-teasers: *www. sandlotscience.com*

- Skeptic's Dictionary: Pareidolia: *http://skepdic.com/pareidol.html*

- University of Colorado at Boulder: Dr. Mark Dubin's psychology, illusions, and sensory phenomena neural links plus web page: *http://spot.colorado.edu/~dubin/bookmarks/b/220.html*

- Vision Science (demonstrations): Internet resources for research in human and animal science: *www.visionscience.com/vsDemos. html*

- WikiPedia: (1) *http://en.wikipedia.org/wiki/Category:Cognitive_ biases* (108 pages of A–Z effects) (2) http://en.wikipedia.org/wiki/ Gestalt_psychology (3) *http://en.wikipedia.org/wiki/Pareidolia* (a type of cognitive bias)

Answers to Questions in Procedure, Exercise #1, step #2

2. The different images seen by different people support the theory that "seeing" involves both selective perceptual filtering and active cognitive processing. It can happen at a very fast, unconscious level that is likely influenced by the viewer's prior knowledge and experiences and/or current state of mind. Seeing is far from a passive, high-fidelity recording of a fixed, external reality.

Activity 6

Utensil Music: Teaching Sound Science

Expected Outcome

The middle portion of a string is tied to the handle of a metal utensil. When the two ends of the string are used to swing the utensil up against a hard surface, the collision generates an audible, but unimpressive vibration. However, if the two ends of the string are brought up against a person's two ears and the collision is repeated, a rich symphonic mix of sounds is easily heard.

Science Concepts

Sound travels better through solids (e.g., a taut string) than through air because sound is a mechanical, compression wave that is transmitted via collisions between adjacent molecules. Unlike electromagnetic waves, sound cannot travel through a vacuum. Also, in contrast to our vision where different wavelengths of light are blended to produce the full spectrum of colors, our hearing enables us to separate complex sounds into their component pitches so that we can dissect and discern a mixture of sounds from different sources.

Science Education Concepts

This simple hands-on exploration highlights the fact that hearing, like all human sensory systems, has natural limitations. Even when we try to be fully attentive, human sampling and processing rates and reception and sensitivity ranges capture only a small portion of the sensory data available in the environment at any time.

Perceptual limitations are another contributing factor to some common, persistent misconceptions that run counter to imperceptible (imperceptible, that is, to unaided human senses) yet valid science concepts, such as the nanoscale world of atoms and molecules. This activity is an *auditory participatory analogy* that broadens and refines students' perceptions, conceptions, and appreciation of "reality." Furthermore, it motivates students to stop and look at, listen to, question, and explore with all of their senses (and technological aids) all that nature and their teacher has to teach them.

Materials

- To prepare a musical utensil chime for each pair of learners: Tie single pieces of ~60 cm string to either the handle end of old utensils (forks and spoons are safer than knives) or the hook of a metal coat hanger (which is *not* as safe as forks and spoons). The two ends of the approximately equal lengths of string (~30 cm each) are left free. The string can be wrapped around the utensil for storage.
- For the Extension activities, use tin cans and paper or yogurt cups and nylon thread to make a homemade nonelectrical telephone. Use Slinkies and Newton's Cradles for macroscropic models of sound transmission.

Points to Ponder

There cannot be mental atrophy in any person who continues to observe, to remember what he observes, and to seek answers for his unceasing hows and whys about things…. Don't keep forever on the public road, going only where others have gone. Leave the beaten track occasionally and dive into the woods. You will be certain to find something you have never seen before. Of course it will be a little thing, but do not ignore it. Follow it up, explore around it; one discovery will lead to another, and before you know it you will have something worth thinking about to occupy your mind. All really big discoveries are the result of thought.

—Alexander Graham Bell, Scottish-American scientist and inventor (1847–1922)

The most exciting phrase to hear in science, the one that heralds new discoveries, is not "Eureka!" ("I found it!") but "That's funny…."

—Isaac Asimov, Russian-born American scientist and science fiction writer (1920–1992)

Procedure

(See answers to questions in steps #1 and #2 on pp. 70–71.)

1. Instruct the learners (working in groups of two who alternate using the "chime") to hold the two free ends of the string between their two thumbs and index fingers and swing the object so that it bumps up against their desk (or have a partner tap the suspended utensil with another utensil). Have the learners (a) write at least three adjectives to describe the sound they hear and (b) brainstorm other instances where sound is associated with the vibration of matter. Ask: Does all sound result from the vibration of matter? Do you think we are hearing all the sound that is being produced? If not, how could we better direct the sound energy to our ears?

2. As one possible answer to the last question, have the learners repeat step #1, except this time press and hold the two free ends of the string up against the middle of their outer ears. Have them describe the sound they hear and compare the experience to the previous one. Ask: Did the sound produced really change from the first to the second trial? If the sound produced didn't change, how can we explain the change in our reception and perception of the sound? What other variables in this setup could be changed and how would you predict that changes in these variables would affect the perceived sound?

Debriefing

When Working With Teachers

This hands-on exploration and the quotes from Bell and Asimov can be used to identify and discuss the key attributes of sound science teaching. At the experiential level, teachers can (1) create engaging, aesthetically pleasing experiences that prompt students to feel a sense of wonder and (2) expand students' conceptions of the reality beyond the limited realm of our natural, unaided sensory perceptions. "Sound science" is more than just a play on words; it is a powerful metaphor for the mind-opening and mind-expanding effect of integrated curriculum, instruction, and assessment.

When Working With Students

This activity could be used either in a unit dealing with sound and hearing or as an introduction to the nature of science and the role of experimentation and technological advances in expanding our perceptions and conceptions of reality. Additionally, in life science classes, attention can focus on the different perceptual worlds that different species inhabit and the selective advantages that these differences provide to allow species to occupy unique niches. Specifically, human hearing is limited to sounds with frequencies between 20 and 20,000 Hz (or ~10 octaves on a music scale) and amplitudes of 0 to 85 decibels. (*Note:* Prolonged exposure to 85 dB or even short-term exposure to sounds above 130 dB causes permanent damage.) These frequency and volume sensitivity ranges are reduced with age and occupational or recreational hearing damage.

Animals' sensitivity ranges are different from those of humans (e.g., bats and whales hear in the ultra- and subsonic ranges of humans). Other species may experience a different auditory reality than we do (e.g., consider the concerns about Navy sonar disrupting the worldwide communication network of whales). This difference between humans and other species is also true, for example, with respect to the sensing of electromagnetic frequencies (*sight*— e.g., plant-pollinating bees and birds see in the ultraviolet and pit vipers see in the infrared and *odors*—e.g., dogs can be trained to smell bombs or drugs).

Many advances in science and human history have relied on technological inventions that have extended the ranges of our natural abilities to detect and process environmental information (e.g., the use of infra- and ultrasound instruments to detect mechanical vibrations or compression-type, longitudinal waves propagated through matter; these instruments range from earthquake seismographs to ultrasonic imaging of a pregnant woman's womb).

Activities #7 and #13 in this volume also explore sound concepts.

Extensions

1. *Tin Can (or Yogurt Cup) Telephones.* Have six students line up, three in the front and three in the back (left side, middle, and right side) of the classroom. Give each student in the front of the room a different "secret whisper message" to say to his or her partner across the room. Ask each student to say the message into the open end of a tin can (or yogurt cup) "telephone" that is connected (through a small hole on the bottom) via a taut, strong cotton string (or nylon thread, fishing line, or kite line) to another can (or cup) held up to the listening partner's ear. (*Note:* The two ends of the string can be secured in place in the hole in bottom of the cans by large knots, paper clips, or buttons. This activity again demonstrates that sound travels better through solids than through air.)

    Variations include the following:

    1. Test how the tautness of the string affects sound transmission (if the string is not pulled to a minimum tension, sound will not be transmitted).

2. Test the effect of a third student using his or her index finger and thumb to hold the string at the midpoint (sound transmission will stop).
3. Test which type of line works best (metal wire, nylon, cotton, or even a long piece of rubber band).
4. Test whether you can add a third party to the conversation by tying a single-line extension and cup at the midpoint of the initial line.
5. Challenge the students to come up with a way to ensure that the volume of sound produced in the two trials is held constant (e.g., replace a student in the activity with a portable radio).

Ask students application-level questions related to the transmission of sound through solids:

- In movies about scouts in the Wild West, buffalo hunters and railroad workers are seen holding their ears to the ground or rail track. Why did they do that?
- In old movies and TV shows, snoopy neighbors who live in adjacent apartments are sometimes seen trying to listen in on their neighbors' conversations by holding empty glasses up against their shared wall with the bottoms pressed against their ears. Why do they use glasses?
- Scientists who study earthquakes (seismologists) can easily "hear" and "feel" an earthquake that happens anywhere on Earth with instruments mounted to the ground that pick up sound vibrations, even though the scientists cannot hear the sound of the earthquake through the air unless they are quite close. Why is this?

2. *Macroscopic Models for Sound Transmission.* A toy Slinky can be used to model a longitudinal (or compression) wave such as sound to show how the energy is transmitted through the medium without permanent displacement of the originally affected molecules. Similarly, Newton's Cradle (a device where a row of five steel balls are suspended by nylon threads) or a row of marbles arranged along the groove of a ruler can be used to model the molecule-to-molecule transmission of sound energy. The closer the molecules (or marbles) are to each other the more efficient the sound transmission (i.e., sound travels better through

Safety Note

Metal wire (Variation #3 in Extension #1) can be sharp and cut the skin. Handle with extreme caution.

solids than through air). Students can also research the history and science behind Alexander Graham Bell's invention of the telephone up through modern cell phones.

3. *Research and Careers in Audiology* involve the improvement of the quality of the aural reality for individuals born with hearing impairments, people who have suffered hearing loss from occupational or recreational sound pollution, and people who are aging. A burgeoning area of research, audiology involves biology, neurology, and mechanical and electrical engineering. The question "Do you hear what I hear?" has become "How can I help you hear what I hear?" Students can use the internet to study and report on this exciting research.

Internet Connections

- American Educator (see also: *www.danielwillingham.com* for other articles and videos): "Brain-Based" learning: More fiction than fact: *www.aft.org/pubs-reports/american_educator/issues/fall2006/cogsci.htm* and Do visual, auditory, and kinesthetic (VAK) learners need VAK instruction?: *www.aft.org/pubs-reports/american_educator/issues/summer2005/cogsci.htm*

- Arbor Scientific's Cool Stuff Newsletter: *www.arborsci.com/Cool Stuff/Archives3.aspx*

- Dallas Symphony Orchestra: Tin can telephone: *www.dsokids.com/2001/dso.asp?PageID=100*

- HowStuffWorks: Homemade toy telephone: *http://science.howstuffworks.com/question410.htm*

- Learning and Teaching Scotland: Science of sound animations (8): Speed of sound and sound waves in solids and gases: *www.ltscotland.org.uk/5to14/resources/science/sound/index.asp*

- PhET Interactive Simulations: Sound ("see" and adjust the frequency, volume, and harmonic content of sound waves): *http://phet.colorado.edu/simulations/sims.php?sim=Sound*

- Philomel Records: *www.philomel.com/index.html*. (This site sells two CDs of audio-based illusions: "Phantom Words and Other Curiosities" and "Musical Illusions and Paradoxes," both by Diana Deutsch.)

• University of Iowa Physics and Astronomy Lecture Demonstrations [select Acoustics]: *http://faraday.physics.uiowa.edu*

• University of Virginia Physics Department: (1) eight sound activity stations: *http://galileo.phys.virginia.edu/ outreach/8thGradeSOL/SoundStationsFrm.htm* (2) Fetal ultrasound: *http://galileo.phys.virginia.edu/outreach/8thGradeSOL/Ultrasound Frm.htm* (3) Doppler effect: *http://galileo.phys.virginia.edu/ outreach/8thGradeSOL/SoundfreqFrm.htm*

• Wake Forest University Physics Department: Sound demonstration videos: *www.wfu.edu/physics/demolabs/demos/avimov/bychptr/ chptr6_sound.htm*

• Whelmers #67: Bells in Your Ears (variation on the Utensil Music activity [Procedure, step #1] using a pencil): *www.mcrel.org/ whelmers/whelm67.asp*

• Wikipedia: (1) Sound: *http://en.wikipedia.org/wiki/Sound#Perception_ of_sound* (2) Infrasound: *http://en.wikipedia.org/wiki/Infrasound* (3) Ultrasound: *http://en.wikipedia.org/wiki/Ultrasound*

Answers to Questions in Procedure, steps #1 and #2

1. The perceived sound will appear to be a low-volume, nonmusical, cheap or tinny noise when the string is held away from the ears. All sound involves the transmission of vibrational energy via molecular collisions.

2. The perceived sound is a louder, fuller, almost orchestral sound when the string is held up against the ear. Both the volume and quality (or timbre) are greatly enhanced. The same sound is being produced in both cases. In the first case, part of the sound produced traveled to our ears through air and part was dissipated through vibrations in the string, which were absorbed by our fingers. In the second case more of the energy traveled directly to our ears through the taut string. Sound vibrations travel better in solids than in air. This phenomenon can be related to the relative closeness of the particles in solids versus gases as visualized by the kinetic molecular theory.

Students could explore the effect of changing experimental variables such as the type of string (e.g., cotton, nylon, or thin metal wire) and the length of string. They could also vary the material or the type of utensil (e.g., plastic, wood, or metal; spoon or fork). They also might measure the length of time different students can hear the sound when a string-and-utensil setup is released from a constant angle (there will be some biological variation with hearing sensitivity).

Activity 7

Identification Detectives: Sounds and Smells of Science

Expected Outcome

The topic of sound is introduced via a mixer activity that involves finding one to three people in the class who have the same "black box system"—an opaque plastic canister—based on how the canister sounds (and feels) when shaken. A similar activity can be done using the sense of smell (see Extensions).

Science Concepts

Sound energy is produced as a mechanical vibration that is transmitted as a longitudinal or compression wave from particle to particle through a medium. The activity serves as a concrete example of and *visual participatory analogy* for the work of scientists who sometimes encounter black box systems that they cannot open and peer inside (given current technologies). In such cases, they make creative leaps of imagination to infer internal composition based on indirect observations and measurements (i.e., empirical evidence), logical arguments, and skeptical review. (See Materials for the optional use of balances to assess students' prior skills with massing objects and to emphasize the quantitative nature of "sound" science. This activity and a related Extension variation on smell can be used in life science classes to begin a unit on the evolution and selective adaptive advantages of sensory perceptual systems.)

Science Education Concepts

This simple, hands-on exploration can be used to emphasize the role of sensory perception as a necessary antecedent to mental conception, the nature of science, and the pedagogical value of a "fun phenomena first" or "wow and wonder before words" instructional strategy. As a mixer, it has students form learning groups by identifying other students with the same "black box." The mixer is a good introduction to a unit on sound. While most hands-on explorations emphasize the sense of sight, this one depends on sound and touch (or smell). It also demonstrates that musical and bodily-kinesthetic intelligences are differentially distributed among students.

Materials

- 1 35 mm canister or other comparably sized, plastic opaque bottle for each learner. Digital photography is making it harder to obtain free 35 mm canisters from photography stores. They may be purchased from science suppliers such as the following:
  - Arbor Scientific. *www.arborsci.com* 800-367-6695. film canister/ PX 1028. $0.75 each

- Educational Innovations. *www.teachersource.com* 888-912-7474. surplus film canisters/CAN-150. $12.95 (~40 canisters)

Alternatively, clean, empty, opaque, nonprescription pill bottles of uniform size will also work.

- *Optional:* If desired, balances can be used to determine the mass of the canisters as an additional piece of verifying information on the match between any two samples. Also, magnets could be made available to determine if any of the hidden contents are magnetic. Water containers can be used to determine the containers' floating orientation based on density. It is also possible to assemble two sets of canisters that have the same mass (and perhaps even sound similar) that would have different magnetic properties. This simple activity can be made just a little more complicated than shaking a wrapped birthday gift to guess its contents or as challenging as a model-building laboratory experiment involving multiple items per canister or substances that change when shaken (e.g., a small quantity of liquid soap water that reversibly forms gas-filled soap bubbles).

Points to Ponder

So little done, so much to do [his last words].

—Alexander Graham Bell, Scottish-American scientist-inventor and educator of the deaf (1847–1922)

They never will try to steal the phonograph. It is not of any commercial value.

—Thomas Alva Edison, partially deaf American inventor (1847–1931)

Procedure

Pre-Class Preparation: Divide the canisters into groups as large as the desired size of your cooperative learning teams (i.e., two–four). Use number or letter labels to identify each canister so you can keep track of the contents of the canisters without having to remove the lids. Do NOT have matched sets listed in an order identifiable by the learners. Put equal amounts or numbers of identical objects in each group of matched canisters. Use a variety of household materials to make the desired number of sets of canisters. Sample items to use include BBs, buttons, cotton balls, lentils, marbles, paper clips (metal and plastic), pennies, washers, small screws, iron fillings, puffed cereal, pushpins, rice, road salt, sand, sugar, sunflower seeds, and thumbtacks. More careful discernment will be needed if multiple canisters contain different numbers of the same items (e.g., one group has two pennies and another group has three pennies).

1. Randomly distribute a sealed canister to each student and instruct the students to "shake, rattle, and roll" (but NOT open) their "black boxes" to become familiar with the sounds the canisters make when treated in specific ways. Learners then circulate around the classroom to compare the sounds of their canisters with those of their peers to find two or more buddies with containers that sound as if they all contain the same object. Emphasize that the canisters should not be opened at any time.

   When used as a mixer with teachers attending a professional development session: You may want to introduce the activity by asking teachers to share their names, one thing they'd like to learn about, one barrier they face in teaching minds-on science, and how they currently understand the two-way interactive nature of perceptions and conceptions.

   When used with students starting a unit on sound energy: Ask them to share one fact they know and one question they have about the science of sound. If these ideas are written down on index cards and collected, they serve as quick, preinstructional diagnostic assessments.

2. As the groups of students with the same sounds start to coalesce, encourage students to brainstorm indirect ways that they could gather more information about their canisters' unknown contents. Encourage cooperation among individuals who have not found their matches, and, if necessary, use your answer key to help individuals who have not yet found their groups or provide balances to help students confirm the groups to which they belong. If time is available, the teacher can provide a broad range of materials, along with empty canisters, for the groups to attempt to build a model that has the same properties and behaviors as the unknowns in their canisters.

3. Ask the small groups to discuss one or more of the following questions depending on whether the intended focus is on physical science, life science, or the nature of science:

(See answers to the following questions on p. 82.)

   a. Did you find this activity challenging? If so, why?
   b. Does all sound involve the vibration of matter? Can sound travel through a vacuum?
   c. Is it easier to "prove" that two things are identical or that they are different? Why?
   d. How does building a model of the system increase our confidence in our identification?
   e. Why was the sense of hearing crucial to the survival of humans in prehistoric times and why is it also crucial in modern, nonindustrialized societies? How is (or was) the sense of hearing like and unlike the sense of sight? What challenges do the hearing impaired and deaf face in our modern society?
   f. As a social species, what unique advantages do we have when it comes to cooperating?
   g. How are the black box systems used in this activity similar to systems studied by scientists and engineers?
   h. What real-world objects or processes cannot currently "be opened," and what new imaging techniques are allowing us to peer inside some of nature's black box systems?

Debriefing

When Working With Teachers

In addition to discussing any of the questions on page 77, challenge science teachers to consider ways they can increase the signal-to-noise ratio (S/N) of the audio components of their lessons—that is, how can they amplify the impact of their verbal messages to students in the midst of a classroom's background noise and distractions. Professional actors learn to intentionally vary their volume and pitch, as can teachers. (See Tauber and Sargent Mester 2007 for other nonverbal ways of increasing S/N. Also consider teaching students the skills of active listening.)

Our mental constructions (i.e., concepts and conceptual networks) are built from the foundation of our perceptions (i.e., environmental inputs through the selectively biased channels of our senses). In this sense, all observations are theory laden. By starting units with engaging activities that feature "fun phenomena first," teachers generate need-to-know questions that students will want to answer.

When Working With Students

This activity is a simple Engage-phase, hands-on exploration activity to be followed by more experiments on the science of sound (see activities #6 and #13). (Also see Appendix B for a description of the 5E Teaching Cycle, of which the Engage phase is a part.) Challenge students to consider how in our modern, sci-tech world, we are bombarded with sounds from traffic, construction, TV, the radio, and other sources, many of which bid for our attention, seek to alter our moods, get us to buy specific products, attend certain movies, and alert us of potential danger. Share the quotes from Bell and Edison, and challenge students to think about how the growth of scientific knowledge has allowed new generations of technologies to supplant landline telephones and phonographs with cell phones and digital recordings.

Extensions

1. *Odoriferous Olfactory Observations: The Nose Knows.* Learners use their sense of smell (rather than sound) to find their matches in the classroom and attempt to identify unknown substances.

Odoriferous solids or different smelling liquids dropped onto and absorbed into cotton balls can be used in canisters that have had small holes drilled into their tops to allow detection by sense of smell. Sample solids include spices and flavorings such as basil, black pepper, chili powder, cinnamon, cloves, garlic, ginger, ground coffee, mint leaves, nutmeg, oregano, and thyme. If using ground powders, place them in a fine mesh bag inside the canisters (or use the mesh and a rubber band to cover the opening of the canisters before securing the top) to prevent the powder from falling through the drilled holes. Sample liquids include extracts such as almond, banana, lemon, orange, peppermint, and vanilla and/or other liquids such as lavender oil, pickle juice, scented oils, and odorless water.

Demonstrate the wafting technique for smelling the contents of the canisters. If you wish to keep the canisters from cross-contamination for future use, store identical sets in separate plastic containers that are solid and sealed after completing the activity. (*Note*: Many odors can diffuse through plastic bags.)

In biology, a sense of smell was one of the earliest chemical detection systems to evolve. (Organisms notice smells on a need-to-know basis; the evolution of more complex nervous systems was driven by an "economic" reality of a kind of unconscious cost/benefit analysis.) Smell continues to play a major role in the survival of many animals whose abilities to detect as many as 10,000 different odors in very low concentrations greatly exceed those of humans. Plants have co-evolved to use odors to attract animals to aid in the plants' reproduction. Foods, toxins, and sexual mates may be identified by their chemical smell "signatures" and social communication is often linked to pheromones. While the role of pheromones in human sexual attraction is still being studied, perfume and cologne industries vie for consumer dollars with claims of producing the smell that will make the consumer irresistible to members of the opposite sex. (See Internet Connections: SFN, Mystery of Smell, and NOVA Online.)

2. *Termite Trails.* The fact that termites will track a continuous line made with a Bic Papermate pen and other common pens (but not markers, pencils, or simple grooves made in paper) makes for a

nice biological black box system. Students can discover that this behavior is related to smell rather than sight and color (see Internet Connections and the products found in science supply catalogs such as Ward's Termite Trails: Follow the Ink Lab Activity (Cat.# 87 V 3525. $32.50). Termite societies are organized into rigid castes: a few reproductives, a larger number of soldiers, and—the most numerous caste—the blind, sterile, 7–10 mm long workers. Using the worker termites in the classroom eliminates the risk of inadvertently creating an infestation if termites escape.

3. *Black Box Basics and Beyond.* See the Internet Connections for other teacher-built demonstrations and commercially available black box experiments that teach the nature of science and its reliance on empirical evidence, logical argument, and skeptical review. See also commercial applications of nondestructive testing (NDT).

4. *Silent Movies and Remember That Jingle.* Demonstrate the importance of sound in our multimedia world by playing a short video clip from an old silent movie, a modern movie, or a TV commercial with the sound off. Even if close captioning of dialogue is available, much of the emotional power and impact of a scary or action-packed scene or a great commercial is lost when the sound is muted. Challenge students to see how many ads, TV shows, or movies they can remember by the jingle, song, or instrumental music that accompanies them. Also, students can be taught to rewrite lyrics for popular tunes to help them use their musical memories to remember scientific information. Under Internet Connections, see the website Neuroscience for Kids for examples of "brain songs" that emphasize the idea of neural connections and meaning-making as linked to familiar tunes.

5. *Historical Connections.* Students can research the biographies of Bell and Edison, who were contemporaries of each other. Bell was a professor of vocal physiology and a teacher of the deaf; Edison, in addition to being a brilliant inventor, had profound hearing loss for most of his life (something he viewed as an advantage rather than a disability!). Both were highly inventive men who saw science as a means to improve the human condition. Edison even invented the idea of an "invention factory" to

80

systematically do research and development with the intention of creating commercial products (e.g., moving pictures and the phonograph) that would serve humanity and generate a profit. Still, as the quote shows, he underestimated the value of his phonograph in creating a worldwide music industry.

Internet Connections

- American Educator (see also *www.danielwillingham.com* for other articles and videos): Do visual, auditory, and kinesthetic (VAK) learners need VAK instruction?: *www.aft.org/pubs-reports/american_educator/issues/summer2005/cogsci.htm*

- Doing Chemistry (movies of demonstrations): Mystery boxes: *http://chemmovies.unl.edu/chemistry/dochem/DoChem001.html*

- International Mind Brain and Education Society: *www.imbes.org*

- Lab-Aids Inc.: Ob-Scertainer® A Better Black Box, Kit No. 100, $69.95/24 students: *www.lab-aids.com/catalog.php?item=100*

- Magic Water Black Box Activity: *www.scienceteacherprogram.org/genscience/Chien05Lesson/INDEX.HTM*

- Mystery of Smell: The vivid world of odors: *www.hhmi.org/senses/d110.html*

- Neuroscience for Kids: Brain songs: *http://faculty.washington.edu/~chudler/songs.html*

- Nondestructive Testing (NDT) Resource Center: Commercial applications of NDT: *www.ndt-ed.org/AboutNDT/aboutndt.htm*

- NOVA Online: Mystery of the senses (five-part series with naturalist Diane Ackerman): *www.pbs.org/wgbh/nova/teachers/programs/22s2_smell.html*

- Society for Neuroscience (SFN): Brain facts: A primer on the brain and nervous system: 74-page book, CD, and free pdf file: *www.sfn.org/index.aspx?pagename=brainfacts*

- Termites, Ink Pens and Pheromones: *www.learnnc.org/lessons/JackiClark5232002016*

- Termite Trails: *www.uky.edu/Ag/Entomology/ythfacts/resourc/tcherpln/termtrails.htm*

- Virginia Tech Physics Lecture Demo W20: Buzzer in a vacuum: *www.phys.vt.edu/~demo/demos/w20.html*
- Ward's Natural Science Co: Black Box Kits: #15 V 9878, 44 boxes and materials: $79.95: *www.wardsci.com/product.asp_Q_pn_E_IG0003323_A_Black+Box+Experiment*

Answers to Questions in Procedure, step #3

a. Most humans are used to relying heavily on their sense of sight. The restriction of not being able to open and look inside the canisters forces the learners to rely on their senses of hearing and touch to guess the identity of the canisters' contents.

b. Sound involves the vibration of matter via molecule-to-molecule collisions. Sound cannot travel through a vacuum, despite what students have seen in outer-space science fiction movies. This can be demonstrated live or via video clips (see Internet Connections: Virginia Tech).

c. Short of opening the canister, one cannot prove conclusively that the contents of two canisters are identical (and even then our sense of sight could be tricked). We can only say that in all the tests performed, the two canisters behaved identically. It is easier to determine that two canisters are different, as only one falsification test is needed. The logic of probable truths is important in both science and courts of law (the idea of "guilty beyond a reasonable doubt").

d. An operational model of a black box system that behaves in the same fashion as the unknown system lends confidence to the provisional truth of our informed guess. Scientists use mental, mathematical, computer simulation, and physical models to extend our knowledge.

e. Like stereoscopic vision, binaural hearing (that is, having two ears) enables organisms to assess the identity and relative location of a wide variety of environmental sounds and determine if the other organisms represent friend, foe, or food or are not immediately relevant. Hearing has an advantage over sight in that it works in the dark and around physical obstructions that limit seeing. Hearing impairment can result in psycho-social-emotional and cultural isolation, learning challenges, and safety risks. Research on improved hearing aids and prosthetic implants holds great promise.

f. As part of our higher cognitive functions, humans have developed language to communicate and cooperate with our peers. This use of language enables us to move beyond random trial and error and to share our own lessons learned with others. This ability is crucial to the evolution of science.

g. Black box systems are objects or processes that cannot be directly opened up and peered into to discover their inner structure and function. The history of science and technology is, in part, a story of the evolution of means of probing more deeply into previously hidden systems.

h. Ever-more-powerful sound, electromagnetic, and nuclear technologies and telescopes, microscopes, and medical diagnostic probes (e.g., CAT scans, MRIs, and PET scans) are enabling us to open up and look inside complex systems such as black holes, the nanoscale world of atoms and molecules, living human brains, and the human genome, often in nondestructive, noninvasive ways.

Section 3:
Nature of Cognition and Cognitive Learning Theory

Two-Balloon Balancing Act: Constructivist Teaching

Expected Outcome

A smaller, human-breath-filled balloon inflates an identical, but larger balloon (already partially inflated) when a clamp sealing a connection between the two balloons is opened. This outcome is counterintuitive for most observers who believe the misconception that bigger is always better (or more powerful).

Science Concepts

This activity explores LaPlace's law, which is as follows: Membranes exert pressure on their contents that decrease as the membranes are stretched to larger sizes and their thickness decreases. The internal pressure of a spherical balloon is inversely proportional to its radius; as a balloon is inflated and its size increases, the internal pressure of the trapped gases decreases. The combined mass (but not volume) of the gases is conserved in this closed, two-balloon system. Biological-medical applications can be explored. This system also provides an opportunity to investigate the nature of science via a cycle of predict-observe-explain.

Science Education Concepts

This activity can be used as a *visual participatory analogy* to challenge the pedagogical misconception that teaching is primarily about a teacher (i.e., the bigger balloon with more knowledge) using pressure to force his or her knowledge into the passive, less knowledgeable student (i.e., the smaller balloon). Beyond the misplaced metaphor of knowledge as a fluid that can be transferred, teachers have as much to learn from their students as their students have to learn from them. Learning and teaching are interactive, collaborative processes in which both the teacher and learner are changed. Unlike the gaseous mixture in this activity, the knowledge gained by one does not come as a result of the other's loss. In fact, total knowledge is not conserved, but increases!

Familiarity with formal scientific terminology does not necessarily equate to understanding the underlying concepts. Knowledge represented in books and translated into teacher talk cannot be transferred to mentally passive students. Conversely, it is possible for those students who lack formal scientific terminology or an understanding of underlying science concepts to possess a solid experiential base that can lead to correct predictions. Students' prior knowledge and experiences need to be considered (and intentionally activated) when designing and implementing iterative curriculum, instruction, and assessment plans. Constructing new understanding on the foundation of previous learning (and/or reconstructing previous conceptions to accommodate new, discrepant experiences) requires active, minds-on processing rather than passive, minds-off (mindless) memorizing of scientific terminology.

Materials

(*Safety Precautions:* **Students who have allergic reactions to natural rubber latex should avoid contact with latex balloons, gloves, rubber bands, and other latex products.** For more information, see *www.mayoclinic.com/health/latex-allergy/DS00621.* See Extension #1 for a variation that uses soap bubbles rather than balloons. Also, if you choose to let teams of learners construct their own two-balloon systems, be sure that you do not reuse the same balloons in order to avoid the spread of germs.)

Safety Note

Use indirectly vented splash goggles with soap bubble option.

- 6 to 9 pairs of 6–9 in. round balloons that are differently colored but identical in uninflated size and thickness
- Each pair of balloons needs one connector—either an empty, single-hole, 2–3 cm diameter, wooden sewing-thread spool or an approximately 10 cm piece of sturdy Tygon tubing. The latter can be purchased from local hardware stores or science supply companies, which sell connectors with a straight, polypropylene tubing connector on each end (e.g., Flinn Scientific. *www.flinnsci.com* AP8878. $2.70 each).
- 1 or 2 clamps are needed for each of the two-balloon systems (unless the inflated balloons are twisted and held closed manually). Clamp options include strong, spring-closing, alligator clothespins or standard laboratory tubing clamps (e.g., Hoffman-type clamps or Day or Mohr pinchcock styles run between $2 and $3 each at science supply companies). Pairs of cylindrical balloons (rather than round ones) can be used as a variation.

Points to Ponder

Nothing has such power to broaden the mind as the ability to investigate systematically and truly all that comes under thy observation in life.

—Marcus Aurelius, Roman emperor and philosopher (121–180 AD)

If you want an answer from nature, we must put our questions in acts, not words, and the acts may take us to curious places.... That is one of the things I like about scientific research. You never know where it will take you next.

—John Haldane, English geneticist (1892–1964)

The best way to have a good idea is to have a lot of ideas.

—Linus Pauling, American chemist (1901–1994)

Procedure

Preparation of the Two-Balloon System: If the balloons are new, pre-stretch them so that they inflate fairly easily. Then inflate all the balloons of one color to the same approximate size (e.g., 10 cm diameter), twist and clamp the ends, and affix each balloon to one of the connectors. Repeat this process with the balloons of a second color, except inflate them to a larger size (e.g., 17 cm diameter) and attach them to the other end of what is now a two-balloon system. The two-balloon systems should be secure enough to be examined (without leakage of air) before the twists or clamps are removed.

(See answers to steps #1–#3 and #5 on p. 94.)

1. Give each team of four learners one of the two-balloon systems (or if desired and time permits, show them how to construct one). *When Working With Teachers:* Ask: If the two-balloon system were to be considered as a *visual analogy* for a teacher interacting with a student, which balloon should represent each role (and why)? What are some of the limitations of this model?

2. *Predict Stage.* Ask learners to complete a cycle of think-write-pair-share. First individually, and then in pairs, they brainstorm a range of possible, plausible outcomes about what will happen when the clamp separating the gases in the two different-size balloons is removed. Consider what will stay constant and what will change and what measurements might be useful.

3. Ask the pairs to regroup into teams to discuss what they think is the most probable outcome from their combined lists of possibilities and to develop scientific rationales (i.e., logical arguments) for their choices. For an added (though irrelevant) variable, consider holding the two-balloon system vertically (rather than horizontally) and asking the students if doing that would matter (i.e., will gravity have a differential effect on the higher versus the lower balloon?).

4. *Observe Stage.* Have the learners simultaneously release the twists and/or clamps on their two-balloon systems and note the results (i.e., the empirical evidence is that the smaller balloons typically shrink in size as they lose air to the larger balloons that become even larger. If pressure is applied to the larger balloon by squeezing it, you can temporarily enlarge the smaller balloon, but it immediately shrinks back as you release the imposed external pressure. By contrast, if external pressure is put on the smaller balloon it will rapidly lose air irreversibly to the larger one.).

5. *Explain Stage.* The outcome of the small balloon inflating the larger one will be a discrepant event for most learners. Yet, most learners have prior experience that, if probed, would have led them to a correct prediction (even without formally knowing the underlying scientific concepts). Activate the learners' prior knowledge (and skeptical review) by posing the following questions:

- How many of you have blown up balloons before?
- Is a balloon more difficult to inflate initially, when it is small, or as it gets larger?
- How does this prior knowledge relate to this activity?
- Why did you fail to connect your prior experience to this situation?
- How do our prior experiences with nature help us understand and apply science concepts?

- Do our everyday prior experiences with nature ever seem to conflict with scientific ideas? If so, why?
- What scientific principles correctly predict and explain this "fun" phenomenon?

6. Consider launching another predict-observe-explore cycle that is student initiated with open-ended questions such as, What additional questions do you have about this phenomenon? What additional experiments would you like to design and/or what resources would you like to use to answer your questions? How can we discover some of the real-world applications of this phenomenon?

Debriefing

When Working With Teachers

Discuss the importance of activating and diagnostically assessing students' prior knowledge as a launching pad for learning. This prior knowledge will likely include a mix of valid experience-based, foundation-building conceptual precursors *as well as* misconceptions that need to be challenged.

Use one or more of the quotes on page 89 to elicit discussion on the importance of (1) regular use of interactive, minds-on activities that challenge students to examine everyday phenomena through the lens of scientific inquiry and (2) beginning with open-ended exploration of plausible outcomes (e.g., brainstorming) rather than with giving students the "right" answer (i.e., brainwashing). In light of the counterintuitive outcome of the activity, discuss the multiple problems with the analogy of knowledge-as-a-fluid that teachers possess and "transfer under pressure" to students' brains. The predict-observe-explain approach offers a simple structure for iterative cycles of analyzing discrepant-event activities and encouraging students' minds-on construction of meaning. (*Note:* Posing and solving scientific questions related to interesting phenomena is analogous to being engaged by the mystery stories of Edgar Allen Poe and other mystery writers.) Teachers may wish to review the research on student misconceptions about gases (e.g., see Driver et al. 1994, chapters 9 and 13, for an introduction and overview). For a broader study of misconceptions and their pedagogical implications, see the extensive list of references provided in Activity #3, Extension #6, in this book.

When Working With Students

Discuss the possible influence of uncontrolled variables, such as using one balloon that has been previously inflated and pre-stretched much more than the other balloon. Check to see whether this affects outcomes. Discuss and further explore the scientific principles underlying this phenomenon and its real-world biomedical applications.

Extensions

1. *Double Trouble With Bubbles.* Soap bubbles can be used instead of balloons as a variation of the same experiment. (Either a commercial or homemade solution can be made to last longer with the addition of a small amount of glycerin.) With an assistant, dip one end of each of two, single-hole, empty, sewing-thread spools into a soap bubble solution. Blow a different size bubble through each spool, cover the holes with index cards, bring the two spools together, and remove the cards. Although the soap bubble system does not experience the occasional "failure" or "malfunctioning" of the two-balloon system (i.e., if the strength of the two balloons is in fact quite different), the bubbles variation does not allow time for discussion of possible outcomes after seeing the initial setup and before the bubbles pop.

   The bubble system is quite intriguing and beautiful, and if long-lasting soap bubbles are used and a light is shown on the bubbles in a darkened room, an interference pattern of shifting colors is observed until the expanding larger bubble turns black and pops. The fact that a soap membrane exerts pressure on the air inside can be demonstrated by blowing a bubble on a straw or funnel and directing the open end against a candle flame. For more information on soap bubbles, see Internet Connections: Exploratorium. Students might also try a two-water-balloon variation on this investigation.

2. *Biomedical Applications.* LaPlace's law explains the collapse of arterial and capillary walls in aneurysms, dilated cardiomyopathy (typically in adults), and the breathing difficulties of premature babies. Use the Internet Connections to explore these and other applications.

3. *Mathematical Measurement Merriment.* Challenge the learners to devise ways to investigate whether the total mass, volume, and/or surface area are conserved in this closed system.

Internet Connections

- Bubbles, Babies and Biology: The story of surfactant (article from the FASBE [Federation of American Societies for Experimental Biology] Journal): *www.fasebj.org/cgi/content/full/18/13/1624e*

- Dilated Cardiomyopathy: *http://library.thinkquest.org/C003758/Function/laplacelaw.htm*

- Exploratorium: Bubbles: *www.exploratorium.edu/ronh/bubbles/bubbles.html* and Bubbles float on CO_2: *www.exploratorium.edu/snacks/bubble_suspension/index.html*

- LaPlace's Law: *http://hyperphysics.phy-astr.gsu.edu/Hbase/ptens.html#lap*

- LaPlace's Law and Lung Misconceptions: *http://advan.physiology.org/cgi/content/full/27/1/34*

- Lung Functioning and Gas Laws: *www.anaesthetist.com/icu/organs/lung/Findex.htm#lungfx.htm*

- Nanopedia: Soap bubbles as nanoscience: *http://nanopedia.case.edu/NWPage.php?page=soap.bubbles*

- Wikipedia: LaPlace's law and medicine: *http://en.wikipedia.org/wiki/Young-Laplace_equation*

Answers to Questions in Procedure, steps #1–#3 and #5

1. In this analogy, teachers are likely to suggest that the teacher is the "bigger balloon" due to his or her greater amount of science knowledge relative to that of the "smaller balloon" students. Conversely, someone might suggest that the smaller balloon represents a state of knowledge that is more compact and organized in teachers versus the less dense and organized state of knowledge in the students.

   In any case, the analogy is flawed in a number of ways that contrast older behaviorist and newer constructivist models of learning: (a) knowledge is not a fluid that can be transmitted or transferred from a teacher to a learner, (b) one individual does not have to lose knowledge for another to gain it (although teachers may feel "exhausted" at the end of the day), (c) learning is not a passive reception process nor is teaching a process of imposing greater "pressure" on the learner than the learner exerts on the teacher, and (d) in the best educational environments, everyone is both a learner and a teacher.

2. Plausible outcomes include (a) the two balloons will remain the same size with no visible change, (b) the larger balloon will transfer air to the smaller one until they are the same size, (c) the smaller balloon will become deflated as it inflates the larger balloon, (d) air will alternately cycle between the balloons, (e) the balloons will shoot off in opposite directions, and (f) the combined pressure will pop one or both of the balloons. If the diameters of the balloons are measured, their surface areas and volumes can be calculated. One could also measure their mass. If the system remains sealed and no air leaks out, the mass of air should remain constant, but the total balloon surface area and volume will not. It is important to note that measurements matter and that calculations "catalyze curiosity" IF students are first challenged with an engaging context such as a discrepant event that demands explanation.

3. Most learners are likely to predict that the larger balloon will inflate the smaller one until they become equal in size. They will make this prediction based on an intuitive "rule" that bigger is always better, stronger, or more powerful and/or that balance or equilibrium always results in equal or equivalent outcomes. At the high school level and above, concepts such as pressure, volume, equilibrium, conservation, and tension are likely to be used in students' "scientific sounding" explanations.

4. [There are no questions or answers for step #4.]

5. The thickness and surface tension of the smaller balloon are greater, and the smaller balloon is less flexible and harder to stretch or blow up than the larger balloon. As the balloon's size increases, its membrane gets thinner and it becomes easier to blow up. Nearly everyone has experienced this when blowing up a balloon, but this experimental setup does not typically cause those particular memories to be activated in students. Unfortunately, for many students, school science is a distinct academic

subject that remains disconnected from real-world applications. If not critically analyzed, everyday experiences often appear to conflict with scientific laws (e.g., different objects do fall at different rates and objects pushed in a straight line do stop due to the effects of friction; both of these examples do "obey" Newtonian physics if all the relevant forces are considered).

Beyond LaPlace's law, another way to explain why a smaller balloon is easier to blow up than a larger balloon is to consider that fluids in flexible containers will assume a shape with the smallest total surface area. Much of the behavior of bubbles and other membrane-bound systems (e.g., living cells) can be explained, in part, by this effect. In the two-balloon (or two-bubble) system, a single larger-volume balloon (or bubble) has less surface area than do two separate balloons (or bubbles) containing the same total mass of air. Relevant mathematical equations include the following:

$r = C/2\pi$ and $V_{sphere} = 4/3\,(3.14)\,r^3$

and

$SA_{sphere} = 4\,(3.14)\,r^2$ where r = radius, C = circumference, V = volume, and SA = surface area

6. Refer students to the Extensions and Internet Connections as possible starting points for further investigations.

Activity 9

Batteries and Bulbs:
Teaching Is More Than Telling

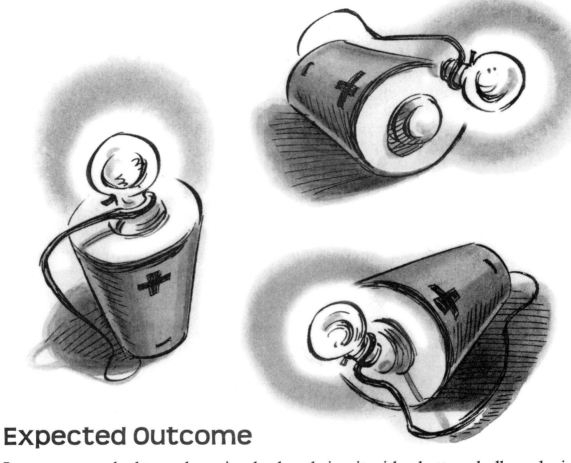

Expected Outcome

Learners are asked to make a simple closed circuit with a battery, bulb, and wire to model the operation of a flashlight. Regardless of their age, many learners experience difficulty with the seemingly simple task of lighting a bulb.

Science Concepts

This simple, three component system can be used to explore a variety of concepts including closed versus open electrical circuits and conductors versus insulators. It also can initiate further explorations into the design of the common direct current (DC) electrical energy storage system and what happens when individual electrochemical cells are connected in series to form a battery.

Science Education Concepts

A major pedagogical misconception that needs to be challenged in teacher education classes and professional development sessions is the default model of "teaching as telling and learning as listening" (i.e., knowledge transmission-reception or "I taught it; therefore, they must have caught it"). Research indicates that even if learning is hands-on, it is still rote learning unless the learner employs active, minds-on construction of concepts (as facilitated by social interactions with a teacher and peers and by direct experiences with phenomena).

In many cases, what passes for understanding is more a matter of being able to use scientific-sounding vocabulary than actually being able to apply knowledge to solve a problem or complete a task. It is also possible to discover how to do some hands-on task by trial and error without developing an understanding of the underlying science concepts. Both types of pseudo-learning can be discussed in the context of this hands-on exploration along with the *visual participatory analogy* of understanding or wisdom as light and a lightbulb as the symbol for bright, creative ideas. At the beginning of the school year, photos can be taken digitally (or with an older Polaroid camera if one can be found) to help the instructor connect the names and faces of the learners. Ask them to think about how science inquiry activities can help "develop" scientific concepts, skills, and attitudes.

Materials:

- A resealable baggie for each team of two learners (or triads if supplies are limited). Each bag contains the following materials:
  - Flashlight bulb (#40 or #41)
  - ~12 cm piece of plastic coated twist-tie with 0.5–1 cm of bare wire exposed at each end
  - 1 standard, inexpensive 1.5 V zinc-carbon battery (AAA, AA, C, or D size; more expensive, longer-lasting alkaline batteries can also be used)
- 1 flashlight (for the teacher-demonstrator)
- *Optional:* Piece of masking tape or transparent tape and hand lenses
- *Optional:* View *The Private Universe: Minds of Our Own.* Program 1: Can We Believe Our Eyes? (Annenberg Foundation 1997, *www. learner.org/resources/series26.html*). Graduates of prestigious universities are shown struggling to light a lightbulb with a battery and wire (an activity normally used in elementary school). This video is available as a free download. Also see other related projects from the Harvard-Smithsonian Center for Astrophysics (e.g., MOSART).
- *Optional:* A song such as "You Light Up My Life" can be played for fun during the hands-on work.

Safety Notes

1. Wear safety glasses or goggles when working with wires and bulbs.

2. Caution students not to touch or work with a "leaking" battery; acid from the battery can burn skin and eyes.

FLASHLIGHT BULB

Points to Ponder

Nature and Nature's laws lay hid in night.
God said, "Let Newton be," and all was light.

—Alexander Pope, English poet-satirist (1688–1744)

Come forth unto the light of things,
Nature be your teacher.

—William Wordsworth, English poet (1770–1850)

Results? Why man, I have gotten a lot of results,
I know several thousand things that don't work
[Edison's response when asked about his numerous
failed attempts at designing a successful lightbulb]....
Many of life's failures are men who did not realize
how close they were to success when they gave
up.... I speak without exaggeration when I say that
I have constructed three thousand different theories
in connection with the electric light....Yet in only
two cases did my experiments prove the truth of
my theory...., Opportunity is missed by most people
because it is dressed in overalls, and looks like work....
Genius is one percent inspiration and ninety-nine
percent perspiration.

—Thomas Alva Edison, American inventor (1847–1931)

Procedure

(The procedure can be done with either teacher-learners or student-learners. Use age-appropriate language and different time allotments for each group.)

1. Begin the activity by turning off all the lights in the room and announcing that before the rapid growth of scientific understanding following Galileo (1564–1642), Newton (1642–1727), and the Age of EnLIGHTenment (mid to late 1700s), theoretical understanding of the physical reality of our world was extremely limited

and in many cases founded more on erroneous superstition than on either scientific hands-on empiricism or minds-on rationalism. Turn on an overhead or multimedia projector to display and read the quotes from Alexander Pope and William Wordsworth as they apply to "natural philosophy" (an earlier name for science).

2. Turn the lights on in the room and mention the analogical imagery of the light of scientific reasoning dispelling the darkness (e.g., Carl Sagan's book *The Demon-Haunted World: Science as a Candle in the Dark* [1995]) and ask learners to identify different electrical means of producing light (e.g., nuclear or fossil-fuel-fired solar, wind, or water-powered AC electrical generators, DC batteries, and fuel cells). If no one mentions a battery and bulb, hold up and use a flashlight. Tell the learners that they are going to explore how a flashlight works as an example of inquiry-oriented, hands-on, and minds-on science.

When Working With Teachers: Mention that well-designed, appropriately guided **hands-on** explorations challenge learners to "hoe" the garden of their own minds to ready the soil for new seeds of understanding. Visualizing this analogy can help teachers focus on the need to ensure that their laboratory exercises result in active, minds-on work with ideas (sometimes referred to as "neurobics") as well as hands-on play with materials.

3. (For answers to the questions in this step, see pp. 107–108.) Distribute a baggie of materials to each pair of learners and ask them to use the battery, single wire, and bulb to build a *model system* of a working flashlight. (*Note:* Two sets of hands are useful to help hold the system together; a piece of tape can also be handy to temporarily secure one end of the wire against one of the two terminals of the battery.) Emphasize that the process of focused investigative play is more important than getting the "right" outcome (lighting the bulb). Learners should draw sketches of both their successful and unsuccessful attempts and begin to develop ideas as to why certain setups work and others do not. That is, what are the common features of all the arrangements that light the bulb? Request that learners (or teams) who quickly achieve the desired outcome refrain from sharing their "right answer" with others; while they are waiting for the rest of the class, they can try to come up with additional arrangements to light the bulb.

4. If available, have the learners use a hand lens to examine the inside of the bulb and the outside of its screw base, display a "dissected," standard-size lightbulb, and/or project an image of the internal design of a lightbulb (such as from Wikipedia or HowStuffWorks; see Internet Connections). The learners should observe that one end of the bulb's filament ends at the bottom of the bulb and the other end is connected to a soldered contact point on the outside of the screw threads.

5. Form teams of four to discuss solutions and to see what is common about all the setups (based on their shared drawings) that allow the lightbulb to light. What is missing in the setups that do not light up? Based on these shared solutions and their observations of the design of the lightbulb, ask the teams to develop one or more preliminary scientific hypotheses that would account for their "successes" versus the "failures." Ask: How do their shared observations "connect" to their theories of what allows the lightbulb to light? Share the Edison quote and briefly discuss how "failures" or "mistakes" can sometimes teach us as much as our "successes."

6. (See comments about this step under Answers to Questions on p. 107.) Flashlights commonly use two (or more) electrochemical cells in a series (to form a true battery). Ask learners to predict-observe-explain what will change if they connect the lightbulb to a two-cell battery (i.e., two cells in a line where the positive terminal of one touches the negative terminal of the other). After they do this, display an image of the 1899 patent for a flashlight that reveals its internal design (see Internet Connections: Wikipedia).

Debriefing

When Working With Teachers

Initiate a whole-group discussion about teachers' experiences with Procedures #1–#6, the "inspirational" quotes from Thomas Edison, and the following open-ended discussion questions:

a. How did the experience feel emotionally and cognitively for those who did not quickly get the "right" answer or never experienced the "Eureka!" experience of success without being shown the setup by a colleague? Successful problem solving depends on affective attributes such as persistence and willingness to play and make "miss-takes" as well as on the learner's prior knowledge and skills. See Extension #1 for an optional DVD on Harvard and MIT graduates and their attempts to light a lightbulb.

b. How can we scaffold inquiry activities so that most students experience an optimal amount of physical and mental challenge (i.e., avoid the boredom that comes from things being too easy or the frustration of things being too hard)? How can we assure our students that "miss-takes" and short-term failures are often necessary for real learning (i.e., that learning involves "fruitful frustration")?

c. How is it possible to have learned all kinds of scientific terminology related to electrical circuits and still have difficulty completing this simple task? Is it possible to be able to complete the hands-on task without having a minds-on understanding of why certain setups work (or don't work)? If so, what does this suggest about the effectiveness of conventional science units that place "answers before questions" and may not focus enough on how (and why) certain lab procedures work to produce desired results?

If you use the camera activity described at the end of Science Education Concepts, point to the digital and/or Polaroid technology as wonderful applications of scientists' integrated understanding of electrical current, light, optics, and digital media (and/or photochemistry). While many common devices are "black boxes" to most users, science classes should at least help students develop some sense of the underlying basics.

In any case, it is important to point out that many seemingly simple systems (such as the one in this activity) involve understanding the subsystems of the various component parts. The system described here has three component parts: (1) the design of the lightbulb, (2) the twist-tie, which consists of an electrically conducting metal wire encased in a plastic insulating sheath, and (3) the battery (which has a somewhat hidden nature that was not directly explored in any detail here).

Follow-up explorations and readings about the battery can include examination of a teacher-dissected battery and/or use images from the Internet Connections (e.g., Wikipedia) that reveal the battery's internal design. The key points to note about the design of a zinc-carbon battery is that it separates the positive terminal (i.e., the metal top end with the raised metal knob that connects directly to an internal carbon rod) from the negative terminal (i.e., the flat metal bottom end, which is connected to an internal zinc metal can that is insulated from the outside of the battery). The actual chemistry is another black box system that requires additional exploration.

Also consider how the incorrect use of language is one source of students' misconceptions about batteries. Technically, individual AAA, AA, C, and D "dry cell batteries" are neither "dry" (if the cells leak and become dried out, they will not work at all) nor "batteries" (but rather single electrochemical cells that should be called *batteries* only when two or more are connected in series). By contrast, a 9 V battery is truly a battery made up of six 1.5 V electrochemical cells stacked on top of each other in series (similar to Volta's original "pile" from the early 1800s). Another misconception related to batteries is the notion that the voltage of a given "battery" is proportional to its size, when, in reality, voltage depends on the battery's chemical composition and the number of separate cells it contains.

The use of the terms *open circuits* and *closed circuits* is also likely to be confusing to students who have heard the common analogy of electrical flow to the flow of water in pipes. They might think incorrectly that an open electrical circuit would be the circuit that would allow electricity to flow and do work. This is an opportunity to point out that some misconceptions are the direct result of incorrect instruction, misuse of scientific terminology in the popular culture (including some textbooks), and the use of terminology that is conventionally correct, but potentially confusing (e.g., open and closed circuits). A large body of research has been published on student misconceptions about electricity (e.g., see Driver et al. 1994; chapter 15 in that book would be a good starting point).

When Working With Students

This classic, inquiry-oriented, hands-on exploration serves as a good Engage- or preliminary Explore-phase activity (depending on the age and prior experience of the students) for a unit on electrical circuits (for a discussion of the 5E Teaching Cycle, see Appendix B). Activity #15 in this book would work as another good Engage-phase activity. Reserve Explain-phase scientific answers until after Explore-phase activities are completed (which would include some explorations related to the design of the battery). Then, during the Explain phase when the teacher introduces formal scientific terms and concepts, students will have the experiential base to make sense of the ideas. Computer simulations are powerful, inexpensive complements and alternatives that can also be used during this phase (see Internet Connections). Later in the Elaboration phase (for high school students) you may want to explore either CFLs (compact fluorescent lightbulbs) or LEDs (light-emitting diodes) as more efficient, economical alternatives to incandescent bulbs and, perhaps, as the new symbol for "bright ideas."

Extensions

1. *Minds-on Multimedia* (for use with teachers). The batteries and bulb experiment dates back to the 1960s National Science Foundation–funded Elementary Science Study (ESS) curriculum project and is featured in the Annenberg Foundation video/DVD: *Minds of Our Own* (see Internet Connections: Annenburg Foundation). It is worth showing the several minutes where university graduates (still in their caps and gowns) are given the same light-the-bulb task as described in this activity. The fact that these highly educated adults—some of whom are shown using scientific terminology apparently without real understanding—are unable to complete this simple, third–fourth grade activity lessens the embarrassment for teachers who have similar difficulty with the task. It also is a great discussion starter on the common educational problem of pseudo-learning that results from a teaching-as-telling, learning-as-listening pedagogical model.

2. *The Three Rs: Reduce, Reuse, and Recycle.* The three Rs can be introduced by using a discarded battery pack from a used Polaroid film pack as an electrical power source and/or lightbulbs cut out of discarded Christmas light strands for batteries-and-bulb DC electricity experiments. (*Note*: Leave 5–10 cm of insulated wire on

each side of a clipped out bulb.) Discarded Polaroid battery packs have a shelf life of years and can be used to allow students to explore parallel and series circuits made from the Christmas lights. In any case, discarded Christmas lights are a cheap, easy-to-use alternative to more expensive screw-in flashlight bulbs. In addition to working with standard batteries, Christmas lights work nicely with handheld generators (see Activity #18 in this book).

Internet Connections

- All About Electricity: Lessons on DC circuits: *www. allaboutcircuits.com/vol_1/index.html*

- Annenberg Foundation: Minds of our own: *www.learner.org/ resources/series26.html*

- Arbor Scientific's Cool Stuff Newsletter: *www.arborsci.com/ CoolStuff/Archives3.aspx*. (See "electricity demonstrations.")

- Electronics for Kids (12 experiments): *http://users.stargate.net/~eit/ kidspage.htm*

- Flash Animations for Physics: Electricity and magnetism: Compare a DC circuit to flow of water: *http://faraday.physics.utoronto. ca/GeneralInterest/Harrison/Flash/#_waves*

- HowStuffWorks: Light bulbs: *http://home.howstuffworks.com/ light-bulb.htm*

- HyperPhysics: DC circuit water analogy (voltage pressure, current-flow rate, and resistance): *http://hyperphysics.phy-astr.gsu.edu/ hbase/electric/watcir.html#c1* (for high school physics)

- Learning and Understanding Key Concepts of Electricity (research article on student conceptions): *www.physics.ohio-stateedu/~ jossem/ICPE/C2.html*

- Lemelson Center for the Study of Invention and Innovation/ Edison Invents: Make a light bulb: *http://invention.smithsonian. org/centerpieces/edison/000_lightbulb_01.asp*

- PhET Interactive Simulations: DC circuit construction kit: *http:// phet.colorado.edu/simulations/sims.php?sim=Circuit_Construction_ Kit_DC_Only*

- A Science Odyssey: Simple circuits, electromagnets and Morse code: *www.pbs.org/wgbh/aso/resources/campcurr/ telecommunication.html*

- Surfing Scientist: Simple circuits and conductivity tester: *www. abc.net.au/science/surfingscientist/pdf/lesson_plan11.pdf*

- University of Virginia Physics Department: Conductors and insulators (how a lightbulb works) and series and parallel circuits hands-on experiments: *http://galileo.phys.virginia.edu/ outreach/8thGradeSOL/Conductors&InsulatorsFrm.htm* and *http:// galileo.phys.virginia.edu/outreach/8thGradeSOL/ SeriesParallelFrm.htm*

- Virtual Voltage Circuit Simulator/Lab: *http://jersey.uoregon.edu/ vlab/Voltage*

- Wikipedia: (1) Flashlight: *http://en.wikipedia.org/wiki/Flashlight* (includes image of the original patent for a flashlight) (2) Incandescent lightbulb: *http://en.wikipedia.org/wiki/Light_bulb* (3) Zinc-carbon battery: *http://en.wikipedia.org/wiki/Zinc-carbon_battery*

Answers to Questions in Procedure, steps #3 and #6

3. Learners with even limited prior experience with the positive (+) and negative (–) ends of a battery and the idea of electron flow should be able to empirically discover by trial and error that if one of the lightbulb's two contact points (i.e., the base that is electrically insulated from the screw threads, which are in electrical contact with the soldered point) is held up against one end or terminal of the battery and the twist-tie wire is used to connect the other contact point to the opposite terminal of the battery, the lightbulb will light. It does not matter which of the two contact points is connected to which of the two ends of the battery. As long as a complete or closed circuit is formed, electrons will flow through the bulb and light it. Three sample arrangements that work to light the bulb are those pictured at the beginning of this activity. (*Note:* The attachments to the positive and negative terminals could be switched and the bulbs would still light.)

6. In a flashlight, the bulb will glow increasingly brighter with each additional cell that is added in series because the voltages of the individual cells are additive. Thus, two 1.5 V cells placed in series will generate 3.0 V. Standard 9 V batteries have six 1.5 cells placed on top of each in series (the teacher can dissect an old 9 V).

Activity 10

Talking Tapes: Beyond Hearing to Understanding

Expected Outcome

A prerecorded auditory message is played back when a "talking tape" is pulled between the thumbnail and index finger and amplified by a paper cup megaphone. The message produced by the talking tape is contrasted to the noise produced by a plastic, hardware store–type tie strip with ridges. The contrast is explained in terms of their different forms and functions. Analogies can be made to biological adaptations and genetic information encoding.

Science Concepts

Sound energy is produced by the mechanical vibration of matter. Information (e.g., via the human voice) can be recorded on a physical medium. Talking tapes are plastic strips with different-size ridges spaced at specific, varying distances. When they are attached to a paper cup megaphone (or taped onto an inflated balloon for sound amplification) and a thumbnail is run along the ridges, a specific phrase is "spoken" (e.g., "Science Is Fun"). Music and language are based on specific patterns of pitches (frequencies), volumes (amplitude), and silence. In contrast, noise has random variations or no variation in these factors.

Thomas Alva Edison was the first person to capture sound waves for reproduction with his original wax phonograph. Later generation vinyl records, magnetic tape, and digital laser discs also "capture" and encode the information from sound waves for subsequent replay. Science analogies can be drawn to biological adaptations and protein manufacturing as encoded by varying nucleotide sequences in DNA and RNA.

Science Education Concepts

Multisensory, hands-on explorations can result in minds-on learning when the teacher scaffolds them with inquiry questions and places the explorations in a well-designed instructional sequence. Understanding requires more than being able to physically hear, pay attention to, listen to, and interpret individual words. Meaning is constructed holistically in light of what the listener-learner already knows and expects to hear. This activity is a visual (and auditory) participatory analogy for the idea that active effort on the part of the learner is needed to decode, amplify (increase the signal-to-noise ratio), and make sense of external stimuli. Hearing with understanding involves more than passive listening (or even hi-fi recording); it requires the active construction of meaning by the listener-learner (see Procedure, step #7). This activity also models how the big picture principles—as seen, for example, in the "Common Themes" of the Benchmarks for Science Literacy and the "Unifying Concepts and Processes" of the National Science Education Standards—can be developed via planned classroom play with simple toys.

Materials

- 24 in. Talking Tapes with prerecorded messages (enough tapes so that each group of two to three has one). Available at novelty stores and science supply stores including the following:
  - Arbor Scientific. *http://arborsci.com.* 800-367-6695. Talkie Tapes. #P7-7320. $16 for 30 tapes (In Internet Connections, see Arbor Scientific's Cool Stuff Newsletter for related demonstrations.)
  - Educational Innovations. *www.teachersource.com.* 888-912-7474. Talking Cups Kit (with instructions): #TC-150. $16.95 for 12 tapes. Also, assorted packages of mixed message tapes: #TC-100 or "Science Is Fun"–only tapes: #TC-100A. $10.95 for 12 tapes.
- Paper cups (or inflated balloons) to act as sound amplifiers (same number as for Talking Tapes) (*Safety Precaution:* **Students who have allergic reactions to natural latex should avoid contact with latex balloons, gloves, rubber bands, and other latex products.** For more inormation, see *www.mayoclinic.com/health/latex-allergy/DS00621.*)
- Plastic tie strips with ridges (same number as for Talking Tapes). Tie strips are available at most hardware stores.
- *Optional*
  - Magnifying lenses or 30× handheld microscopes (preferred for greater magnifying power)
  - Masking tape is used for a biology analogy.

Points to Ponder

What hath God wrought! [First telegraph message sent as a series of dots and dashes in 1844 from a room in the Supreme Court building in Washington, D.C. to Baltimore, Maryland]

—Samuel Morse, American inventor and painter (1791–1872)

Mister Watson, come here, I want you. [First words transmitted via telephone in 1876]

—Alexander Graham Bell, Scottish-American scientist-inventor (1847–1922)

Mary had a little lamb, whose fleece was white as snow! [First words recorded on a phonograph in 1877]

—Thomas Alva Edison, American inventor (1847–1931)

(*Note:* These three quotes represent the first successful attempts to encode and transmit the human voice across distance or time via "new" technologies.)

Procedure

(The answers to the questions embedded in steps #1–#6 will be found on pp. 116–117.)

Introduce the activity by sharing the three quotes (without attribution) and asking, What do these quotes have in common? Have the learners form dyads. Give each dyad a Talking Tape (but do NOT yet identify it as a Talking Tape); a plastic tie strip; and a magnifying lens or a 30× handheld microscope. (*Note*: The handheld microscope produces the up/down and left/right reversal of true microscopes versus the image produced by a single-lens magnifier.) Use the following directions and inquiry questions to guide the learners' hands-on explorations.

1. Examine the two sides of the Talking Tape strip using the senses of touch and sight. Compare this strip with the hardware store plastic tie strip in terms of *constancy* and *change*. Examine the pattern of ridges on the two different kinds of strips using a hand lens (or a 30× handheld microscope if available). What can you say about the relative spacing and nature of the ridges? Describe in words and drawings what you observe.

2. *Predict:* How might you use these two different types of strips to produce sound? Which form of strip do you predict would function as an information system to encode specific sound messages? Which strip would function better as a device for tying and holding? *Observe:* What evidence could you produce for your predictions and claims? *Explain:* Can one design be said to be universally better than the other? Why or why not? *Optional Biological Connection:* How does this principle relate to inter- and intraspecies biological variations, adaptations, and natural selection (or "survival of the fittest")?

3. How might we amplify (or increase the signal-to-background-noise ratio) of the sound produced by the plastic strips? After students brainstorm answers to that question, provide each dyad with two paper cups (and/or deflated balloons) and ask them to see if they can discover how to use the cups as amplifiers (hint: consider a megaphone). If necessary, show the learners how to pull the strip through a slit in the cup and tie off the end (as shown in the drawing and directions provided with the Talking Tape. "Play" the two different kinds of strips with the cup and then without the cup (or inflated balloon). What is the role of the strip versus that of the cup in producing a clearly audible, discernable sound (i.e., the relationship between their different *forms* and *functions*)?

4. What do you predict will happen if you increase the speed at which you run your fingers over the ridges? Then, do it. What other variables can you test?

5. What real-world recording systems does this sound-encoding model remind you of? Why is it important to match the "playback" speed to the original "recording-encoding" speed (e.g., revolutions of 33 1/3, 45, or 78 rpm on phonograph turntables or disc drive speeds on laser disc players)?

6. What do you think would happen to the message if a section of the ridges were covered with masking tape before "playing" the strips or if only one portion of the talking tape is "played"? Try it! *Biological Connection:* How might the idea of covering a portion of the talking tape serve as a large-scale model for gene expression with DNA, RNA, and protein production? As unicellular organisms evolved into multicellular organisms, why was it necessary to be able to block expression of certain genes?

7. *For Teachers Only (an analogy):* How are the original recording and subsequent playback of sound on a Talking Tape, phonograph record, or CD like and unlike the process of learning from a teacher? Is teaching more about indoctrination or inspiration?

Debriefing

When Working With Teachers

The need to combine fun, hands-on "play" with the mentally engaging, minds-on "work" of learning should be emphasized in professional development settings and in science methods courses. In their own classrooms, teachers should also emphasize frequently—in words their students will understand—that learning is a process of active construction of understanding. Students carry out this process by decoding and making sense of external stimuli in light of what they already know. Neither hearing nor learning occurs if the teacher merely sends encoded messages to the passive learner.

Discuss the related analogy that is implied in the questions in Procedure, step #7. Also, interested teachers should be encouraged to explore the research on student misconceptions about sound (e.g., see Driver et al. 1994; Chapter 18 is a good starting point).

When Working With Students

This activity can be part of a physical science unit on sound energy or can be used as a biological analogy for the genetic code in the life sciences (with different questions). Note that in the latter case, the code is more akin to the digital encoding on a CD or DVD laserdisc than the analog encoding on the Talking Tapes (or phonograph records).

Extension

The Evolution of Sound Science. Have students research other sound technologies. Use a hand lens or 30× microscope to see the *analog* form of wavy lines on an old vinyl record. In lieu of an electric turntable and amplifier, students can construct their own alternative by rolling a large piece of poster paper into a cone shape (see Internet Connections: Groovy Sounds). If a needle or pin is inserted through the pointed end of the cone and subsequently allowed to run along the lines of an old vinyl record as it is rotated manually on a revolving tray (such as a lazy Susan), the sound will be audible. Old audiotapes or videotapes will be attracted to a strong magnet and the information on the tape can be destroyed if the tape is pulled across such a magnet. The laser light reflecting patterns on digital CDs are not so easily studied. The evolution from record to audiotape to CD to MP3 formats has been a win-win in terms of both the environment ("lighter" matter demands and less pollution) and enhanced sound quality.

Internet Connections

• Arbor Scientific's Cool Stuff Newsletter: *www.arborsci.com/ CoolStuff/Archives3.aspx.* (See Sound and Waves demonstration listings, including Talkie Tapes.)

• HowStuffWorks: How DNA works: *http://science.howstuffworks. com/dna.htm* and Analog and digital recording: *http://communication.howstuffworks.com/analog-digital.htm*

• Groovy Sounds: *www.exo.net/~emuller/activities/Groovy%20 Sounds.pdf* and *www.exploratorium.edu/snacks/groovy_sounds/ index.html*

• NIH/National Human Genome Research Institute (NGHRI): *www.genome.gov*

• Wikipedia: History of sound recording: *http://en.wikipedia.org/ wiki/History_of_sound_recording* and Gene expression: *http:// en.wikipedia.org/wiki/Expressed_genes*

Answers to Questions in Procedure, steps #1–#6

1. Unlike the evenly spaced, uniform height, continuous ridges on the plastic tie strips, the spacing of ridges on the Talking Tapes is not uniform but seems to form some kind of pattern or code with blank spaces between the separate "words." There is also an amplitude variation within the "words" (i.e., variable height ridges) on the Talking Tapes and angled, "back-catch" ridges on the standard tie tapes. If different talking tapes are used, the tapes that play different messages will be seen to have different lengths, heights, and patterns of ridges and space separations.

2. Sound energy is produced by the mechanical vibration of matter. If either strip is pulled between your thumbnail and index finger it will make a barely audible sound. The specific order and organization of ridges and spaces on the Talking Tapes produce a specific sequence of words when the strips are pulled in the proper direction. The hardware strips have higher ridges (and deeper valleys), and one end is designed to allow the angled ridges to be pulled through and lock in place in one direction only. Even if it had such an end, the talking tape strip would not be able to be adjusted as regularly as the plastic tie strip for tying and securing purposes. One design cannot be said to be universally better than the other; different forms are better suited for different functions or environmental contexts. This fact is analogous to biological variations within and between species that are more or less adaptive for particular environments. The phrase "survival of the fittest" is always context-dependent on the environment.

3. The cup serves as a sound amplifier to increase the volume like a sound box on a guitar. Alternatively, if the strip is pressed up against your ear and you play the tape between your thumbnail and index finger, you will be able to hear the sound without using an amplifier; sound travels better through solids than through air.

4. The rate and pitch of the sounds or words will increase if the tape is pulled more quickly. Students could also vary, for example, the direction in which they pull the tape (i.e., the message will be unintelligible if it is played in reverse) and how much pressure they exert between their thumbs and index fingers.

5. Proper playback speed is needed to match the frequencies of the original message. This also applies to magnetic audiotapes and videotapes and digital recording media.

6. If part of the Talking Tape is covered over with masking tape, part of the message or words will not be expressed during playback. *Biological Connection:* Given the dynamics of mitosis, each somatic cell contains all of the organism's genetic code, but cells "express" only those genes and produce only those proteins specific to their designated mission or function (e.g., skin cells produce skin cells). In contrast to the single strip Talking Tape code, the intertwined, double helix form of DNA allows for "exposure" of select portions of the protein-making template or code. Consider the "digital" nature of genes and the work to map genomes.

7. *For Use by Teachers:* The comparison of the sequence of teaching → learning → testing to a speaker's voice being recorded on a vinyl phonograph record or CD (and then played back for others to hear) is actually a poor, misguided analogy. As a pedagogical model, the sequence of transmission, reception, recording, and playback will, at best, produce short-lived pseudo-learning. Learners' minds are not standard-format, impressible vinyl discs that passively receive and have encoded on them a spoken message that can later be played back with high fidelity. Individual learners uniquely attend to and perceive (or hear) selective parts of the teacher's "message" based on their prior conceptions, their motivation, and various attentional factors, such as how relevant the message seems to them.

  During and following instruction, each learner reconstructs and encodes (rather than records) an altered version of the original "intended lesson." When assessed, learners will vary in their ability to remember, retrieve, "replay," and creatively and correctly apply their reconstructions of the original message. As such, learners' minds can be metaphorically thought of as creative, "reality reconstruction" information-processing workers. Effective science teaching is more about inspiring learners to question accepted truth and push back the boundaries of our individual and collective ignorance. An analogy to rewritable laserdiscs where the learners do their own writing is more appropriate.

Super-Absorbent Polymers: Minds-On Learning and Brain "Growth"

Expected Outcome

A plastic, walnut-size, novelty store toy brain (or animal) expands greatly when left in water over a 24–72 hr. period. When removed from the water and allowed to dry, it returns to its original size. Changes in mass, volume, and density can be measured.

Science Concepts

Hydrophilic, super-absorbent polymers (SAPs) absorb and retain water up to several hundred times their mass and increase in volume up to 600%. These facts can be checked in measurement and graphing skills–based labs. Beyond their use in "grow toys," SAPs are found in disposable diapers, (latex) paint hardeners, nursery potting soils, firefighting gels, and fake snow (in movies and store displays).

Science Education Concepts

Unlike this prop, students' brains do not grow/learn by passive absorption of received knowledge from the outside. *Learning is an interactive, psychologically active, inside-out and outside-in process* that involves both assimilation and accommodation of new experiences and ideas into preexisting conceptual frameworks. These two mental processes—assimilation and accommodation—were identified by Piaget and are analogous to, respectively, filling in missing pieces of a previously partially assembled puzzle in a way that fills in holes or expands the perimeter and to removing and rearranging pieces in a way that produces a different picture. The latter process of accommodation typically involves challenging and resolving misconceptions and mental models of limited applicability. The teacher's integrated curriculum-instruction-assessment plan provides the educational environment that facilitates the development of enduring, yet ever-evolving, neurological networks that are the biological basis of memory and learning.

In professional development contexts that focus on brain science, the Growing Brain can serve as a *visual participatory analogy* to initiate a discussion about brain "growth" as related to learning in educationally enriched environments (see the inquiry questions in the teacher section of Procedure). A linear sequence of increasingly larger Growing Brains can also be used as visual model for the evolution of the brain across species and over evolutionary time.

Materials

Various suppliers (e.g., science education catalogs, science museum and nature stores, novelty shops, and home improvements stores in the paint and garden sections) offer a variety of products made from super-absorbent polymers:

- Creative Presentation Resources. Magic Supplies: Lightning Gel Slush Powder. #604-025. 4 oz. $7.95. *www.presentationresources. net/magic_supplies.html*. 800-308-0399.
- Educational Innovations. *www.teachersource.com*. 888-912-7474. Polyacrylamide products:
  - Growing Brain. #GB-200 (~2 in. diameter walnut-size brain grows to ~5 in.). $3.95
  - Gro-Beast Alligator. #GB-2 (~3 in. grows to ~2 in.). $2.95. Preferred for student measurements
  - Gro-Beast Dinosaurs. #GB-1 (~1 in. grows to ~3 in.). $0.95
  - *Growing Creatures*. Grades 3–8 activity book by Anne Linehan and Betsy Franco. #BK-450. $14.25
  - Oriental Trading Co. *www.orientaltrading.com*. 800-875-8480.
  - Growing Dinosaurs. IN-39/217. 1.75–2 in. 48 assorted dinosaurs. $7.90
  - Fortune Fish. IN-39/761. 144 pieces. $8.99 (for Extension #4 on p. 126)
- U.S. Toy Company. *www.ustoys.com*. Fortune Fish. VL79. 72 pieces. $4.95 (for Extension #4)
- 2 L bottle, empty and with the top cut off (in which to grow the brain)
- 2 L bottle, full size, empty (for a teacher demonstration)
- Plastic tennis-ball can or flat cake pan, empty (in which to grow the alligator; for students to use in conjunction with a linear measurements and graphing skills hands-on exploration)
- *For Extension Activities:* disposable diaper, sponge, and/or gummy bears
- *Optional:* Video clips from sci-fi movies (e.g., on YouTube) that show brains being grown in containers of "vital liquids" could be used for fun when working with teachers.

Safety Notes

1. Wear indirectly vented chemical splash goggles for this activity.

2. Wash hands with soap and water after completing this activity.

3. Super-absorbent polymer powder (Extension #1) is a major irritant and should not be exposed directly to air where students or staff are working.

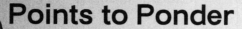

Points to Ponder

You cannot teach a person anything. You can only help him to find it for himself.

—Galileo Galilei, Italian astronomer and physicist (1564–1642)

It is much to be regretted that habits of exact observation are not cultivated in our schools; to this deficiency may be traced much of the fallacious reasoning, the false philosophy that prevails.

—Alexander von Humboldt, German naturalist and statesman (1769–1859)

Procedure

When Working With Teachers (an interactive participatory demonstration)

Pre-demonstration Preparation: Because the process of "growth" in these toys is measured over a period of days, teachers will probably want to pre-soak several samples of the Growing Brain for different lengths of time (one-half day to four days or more). Then, teachers can be shown the variable states of growth in a fashion analogous to what TV cooking show hosts do when they display the transition from a recipe's unmixed ingredients to completed meals in a brief time interval.

In the professional development setting or a science methods class, do the following (answers to questions in steps #1–#3 are on pp. 128–129):

1. Hold up a dry Growing Brain and ask, How is this model brain like and unlike the real thing?

2. *If* the Growing Brain model is like a human brain, *then* we should be able to grow it by placing it in what kind of environment? As all cells are composed primarily of water, place the toy in tap water (as an analog of an enriched educational environment). If you have extra models, you may also want to place one in glucose water for comparison. Also, if available, clips from sci-fi movies of human brains being grown in beakers can be used to elicit laughter about "controlled learning environments and ideal students." Of course, "passive blobs" are the last thing we want our students to be!

3. When no dramatic effect is immediately forthcoming, feign surprise (e.g., "This is not the way the instruction book or the teacher's guide said it was supposed to work!") and continue to discuss the validity of the analogy via the following discussion questions:

- Are the results of enriched educational environments immediately apparent? How long is it reasonable to wait to assess significant changes in learning and cognition?
- Does learning cause the brain to grow larger? If not, how does it change the brain?
- How can teachers measure "brain growth" since we cannot directly observe the brain's neural networks in action? How can assessments create both inside-out and outside-in exchanges between the learner and teacher?
- Do human brains "grow" by passively receiving (or absorbing) and recording (or copying) external stimuli? If not, what are better ways of thinking about learning and teaching?

(*Note:* Be sure to reexamine the immersed brain during the next training session or class. As previously mentioned, for quick demonstration turnaround time, a second or series of prewatered, grown brains can be displayed as immediately available discussion props.)

When Working With Students (hands-on exploration)

The Growing Brains or Gro-Monsters (alligators or smaller, cheaper animals) can be monitored to record, graph, and compare the relative rates of linear, volume, mass, and density growth. For demonstration purposes, a dry alligator can be softened in water for a few minutes and wiggled into the top of a 2 L bottle of water; it will grow to fill the entire bottle! For hands-on explorations, students can measure the growth of one of the animals immersed in water-filled, flat cake pans; in 2 L bottles with the tops cut off; or in plastic tennis-ball cans. Students can graph the growth in one or more of the metric measures over a 1–7 day period and then again for another 1–7 day period as the animal dries out and shrinks. Students should wash their hands with soap after making the measurements and the toy animal should be kept out of direct sunlight, which can degrade the reusable polymer. The *Growing Creatures* activity book (see Materials) describes

14 student projects. The polymer's absorbing power (per equivalent mass) can also be compared to that of a sponge. For a cheaper alternative to the SAP animals, see Internet Connections: Gummy Bear Lab. Unlike the synthetic polymer-based animals, some dissolution and diffusion will occur with gummy bears (i.e., some of the outsides come off and insides come out into the water).

Debriefing

When Working With Teachers

Focus on the analogy-based discussion questions in the extensions (and the answers on pp. 128–129). Teachers may also be interested in facts about the human brain. Today's *Homo sapiens* adult brain is the size of a cantaloupe (1.7 L) with the texture of an avocado. Our brain dates back some 0.1–0.25 millions of years, weighs in at approximately 1.4 kg (3 lbs.), and contains some 100 billion neurons. It can be traced back through its earlier, more modest-size predecessors: the first homonids (2.5 mya) ← first primates (55 mya) ← mammals (235 mya) ← reptiles (320 mya) ← first prokaryotic cells (3.8 billion years ago). Significant increases in the brain's mass, volume, and complexity have occurred during its long history.

When Working With Students

Use hands-on explorations to focus on the skills of measurement and hypothesis testing. In high school chemistry classes, add discussions of the nature and industrial and home uses of super-absorbent polymers.

Extensions

1. *Disappearing Water Trick.* The super-absorbent polymer (SAP) from a disposable baby diaper can be extracted by cutting open the inside panel and shaking the inside of the diaper into a zip-seal plastic bag (i.e., "taking the inside out"). Alternatively, sodium polyacrylate can be purchased from science supply companies such as Educational Innovations and Flinn Scientific (or substitute waste paint hardener; see Extension #3 on p. 126). If a small quantity of the polymer is placed inside an opaque cup (such as a Styrofoam cup), hydrogen bonding will enable

the polymer to absorb a much larger quantity of water poured into the cup (i.e., take the "outside in"). The cup of solid gel will magically appear to be empty when it is subsequently turned over a volunteer student's head.

If you want to heighten the magic show element of this demonstration, use a larger quantity of the polymer and pour one-half cup of water into one of three Styrofoam cups (the one with the polymer). Then ask the students to observe as you play switcheroo with the order of the cups. No matter which cup they guess, when you turn it over, no water will spill out (either the cup is empty or contains the congealed polymer-water combination). You can also explore the negative effect of the addition of salt on the formation of the gel. This science magic trick can lead to science-technology-society (STS) debates and hands-on explorations related to the use of SAPs in disposable diapers (see next step) and industry and agriculture

2. *Diaper Debate.* The topic of disposable diapers versus cloth diapers also makes an interesting STS case study on the relative environmental trade-offs (i.e., energy use versus solid waste and water pollution) associated with natural, human, biodegradable solid waste production. Students can be challenged to develop critical-thinking skills by exploring the latest research on the relative pros and cons of the two kinds of diapers as well as efforts to create a diaper that is more economical and more environmentally and parent and baby friendly. Issues of who is funding and/ or doing the research and potential biases can also be explored. Web-based starting points include the following:

    *   gDiapers (eco-friendly diapers; "g" stands for *green*): *http:// gdiapers.com*
    *   No Clear Winner: (a) *http://abcnews.go.com/Technology/ story?id=789465&page=1* (b) *www.mindfully.org/Plastic/ Diaper-Not-Clear.htm* (c) *www.sustainabilityinstitute.org/dhm_ archive/index.php?display_article=vn321diapersed*
    *   Pro-Cloth Diapers: *www.parentingweb.com/misc/cloth_facts.htm*
    *   Real Diaper Association (reusable cloth): *www.realdiaperas sociation.org*
    *   Wikipedia: Pampers: *http://en.wikipedia.org/wiki/Pampers*

3. *Waste Paint Hardener.* Home improvement stores and paint stores sell SAPs as a means to solidify leftover, waste latex (water-based) paints so that they can be "thrown away" legally via standard curbside solid waste disposal. Students can compare the water-absorbing-capability-to-cost ratio of this product to that of other SAPs bought from science supply companies. They can also explore the pros and cons of this product from an environmental perspective.

4. *Forturne Teller Miracle Fish.* This novelty item is a thin, red colored, fish-shaped plastic film that contains an SAP that causes the fish to curl up when placed in the palm of your hand. Challenge students to figure out the science behind this discrepant event. Most students will probably hypothesize that it is a heat-related phenomenon, analogous perhaps to the unequal expansion that results from heating a bimetallic strip (as found in many thermostats). Additional tests (e.g., placing the fish on the topside of your hand, on a slightly damp sponge, or on a warm dry surface) will show that the key variable is the presence of water (absorbed on the side of the strip that is in contact with moisture), not simply heat. This makes for a simple, low cost, inquiry-based hands-on exploration. Biology students can go on to explore the function (i.e., heat release via phase change) and primary sites of human perspiration.

5. *For Teachers Only: Minds-On Mnemonics.* Challenge small groups of teachers to develop acronyms that summarize what we know about active, minds-on learning using the letters in the word *brains.* Often whimsical (e.g., MOM, created from <u>m</u>inds-<u>o</u>n <u>m</u>nemonics), acronyms can serve as mnemonics to help students remember scientific information. See also Activity #12. Sample mnemonics using the word *brains* are as follows:

<u>B</u>uilding	<u>B</u>egin	<u>B</u>rains
<u>R</u>elationships	<u>R</u>ewiring	<u>R</u>elate,
<u>A</u>mong	<u>A</u>nd (making)	<u>A</u>ssimilate, (and
<u>I</u>nterrelated	<u>I</u>nterconnections	accommodate),
<u>N</u>etworks of	(between)	<u>I</u>deas (when)
<u>S</u>ymbols	<u>N</u>eural	<u>N</u>urtured
	<u>S</u>ynapses	<u>S</u>ystematically

For added levity, edible props that can be used with this Extension are full-size gelatin brain molds and/or bite-size brain candy molds, which allow learners to eat "brain food." These unusual products are readily available through the web.

Internet Connections

- BASF Superabsorbent Polymer Hygiene and Industrial Applications: *www.functionalpolymers.basf.com:80/portal/basf/ien/dt.jsp?setCursor=1_286709*

- HomeHarvest Garden Supply's Soil Moist: *http://homeharvest.com/waterabsorbtion.htm*

- How Stuff Works: Disposable diapers: *www.howstuffworks.com/question207.htm*

- Purdue University Division of Chemical Education: SAP demonstration (QuickTime movie): *http://chemed.chem.purdue.edu/demos/main_pages/25.5.html*

- Steve Spangler's Making Science Fun: Experiments with water-absorbing crystals: *www.stevespanglerscience.com/experiment/00000057*

- Science and Technology Center for Environmentally Responsible Solvents and Processes: Dirty diapers (lab): *www.science-house.org/CO2/activities/polymer/diaper.html*

- Science Spot: Gummy bear lab: *http://sciencespot.net/Media/mmaniabearlab.pdf*

- Watersorb (product for gardeners and farmers): *www.watersorb.com/index.htm*

- Wikipedia: Superabsorbent polymers: *http://en.wikipedia.org/wiki/Superabsorbent_polymer*

Answers to Questions in Procedure (When Working With Teachers), steps #1–#3

1. The toy brain contains polymers and has a textured, folded surface, but it is too small, hard, and dry to be a good model for the human brain. Also, it is not connected to a sensory input and motor output system via an electrochemical network of "wires." Human cognition involves much more than our brains; in some sense, we learn with our whole bodies.

2. The human brain is a physiologically complex, electrochemical "wet-ware, world-processing" system that draws a disproportionate amount of oxygen and glucose to work 24/7. Teachers might suggest that the plastic model should be placed in aerated sugar water!

3. Determining the impact of instructional interventions involves a continuous process of both formal and informal curriculum-embedded assessments that include preinstructional (or diagnostic) assessments, formative assessments, and graded, summative assessments placed at appropriate intervals between different topics and/or units. All assessment can be viewed as processes of getting the "inside out" or taking a snapshot of learners' internal conceptual networks and getting the "outside in" by providing feedback to the learner on the validity of his or her prior conceptions. A key purpose of assessment is to provide feedback that helps both students and teachers adjust, refine, and refocus their efforts to achieve the learning objectives.

    Learning does not cause the brain to grow physically larger, but natural development and physical maturation up through the adolescent years cause the brain to increase in overall size within an expanding skull. At any age, however, learning does lead to the growth and expansion of neural interconnections and rewiring of neural networks. As an active "world processing" system, the brain does not work anything like a passive, mindless, physical recording medium. Learning is always minds-on whether or not it is an intentional or a conscious process (but the more intentional and conscious, the better). Learners also have the means to affect (for better or worse) their learning environments to the extent they choose to be participatory and positive or not (this is in contrast to the Growing Brain, which primarily takes water from its environment without "putting something back"). The best learning environments have teachers and learners alternating both inside-out and outside-in exchanges that enrich everyone's understanding.

Activity 12

Mental Puzzles, Memory, and Mnemonics: Seeking Patterns

Expected Outcome

Various mental puzzles and memory tasks can seem difficult until a heuristic or generalizable problem-solving technique is discovered or invented. For example, a mnemonic (memory device) such as the acronym ROY G. BIV can help learners remember information (in this case, the colors of the rainbow). A heuristic or mnemonic can make an underlying pattern discernable or make a task more meaningful by relating it to prior knowledge.

Science Concepts

The nature of science involves searching for logical patterns in and provisional, tentative "truths" about natural phenomena that hold up to or are modified by subsequent empirical testing and skeptical review. Comprehension and the ability to apply or transfer understanding depend on long-term memory encoding, retention, and retrieval, which are enhanced when learners actively search for and create meaningful patterns. These patterns, in turn, allow for extrapolation and interpolation predictions that lead to further learning, applications, and discoveries.

Science Education Concepts

Learning that lasts is always a psychologically active, minds-on process. Both limited, short-term and unlimited, long-term memory are facilitated by active meaning-making on the part of the learner. He or she connects the new information to prior knowledge and (subconsciously) considers its emotional relevance. Students need to be taught *how* to learn as much as *what* to learn. Guided discovery techniques—where the curriculum-instruction-assessment plan is designed to help learners discover underlying patterns—are especially powerful in reducing cognitive load and bypassing short-term memory (STM) limits.

Research indicates that STM (or working memory) in humans 15 years and older can, on average, retain 7 (+/− 2) chunks of (*unrelated*) information before it is either displaced by new information or transferred into long-term memory (Miller 1956). (Miller's classic article, "The Magical Number Seven, Plus or Minus Two: Some Limits on the Capacity for Processing Information," can be accessed at the Classics in the History of Psychology website: *http://psychclassics.yorku.ca/Miller*.) This principle has pedagogical implications in terms of both the number of new concepts introduced in a given lesson and the extent to which we help students create meaningful cognitive connections by linking new ideas to each other and to prior concepts. In teaching science, the whole should be greater (or make more sense) than the sum of the (unrelated) parts. The four experiments in the Procedure section are not so much analogies as they are direct probes into and models of the processes involved in human memory capacity associated with words and numbers. Learners also have STM limits with respect to visual and spatial information.

Materials

Pattern-type puzzles (see Procedure) are displayed or projected (using a document camera or overhead projector) in a way that they can be subsequently covered or turned off so that the learners' recall can be tested.

Points to Ponder

The art of remembering is the art of thinking…. When we wish to fix a new thing in either our own mind or a pupil's, our conscious effort should not be so much to impress and retain it as to connect it with something else already there. The connecting is the thinking; and if we attend clearly to the connection, the connected thing will certainly be likely to remain within recall.

—William James, American psychologist and philosopher (1842–1910) in *Talks to Teachers on Psychology* (1888)

Of course, intellectual learning includes the amassing and retention of information. But information is an undigested burden unless it is understood…. And understanding, [or] comprehension, means that the various parts of the information acquired are grasped in their relations to one another, a result that is attained only when acquisition is accompanied by constant reflection upon the meaning of what is studied.

—John Dewey, American philosopher and educator (1859–1952) in *How We Think* (1910)

Procedure

Experiment #1: Number Sequences With Mathematical Patterns

(See answers on p. 138.)
Display or project (one at a time) one or more of the following 13-digit number sequences for 10 seconds and ask the learners to see how much of the sequence they can remember without writing it down. After the time has elapsed, remove the number sequence from sight and ask the learners to write down as much of the sequence as they can recall. Tally the results and ask about any "tricks" used by the students who remembered the whole sequence (the statistical outliers).

1392781243729

1491625364964

0112358132134

Experiment #2: Number Sequences and Patterns

(See answers on p. 139.)
a. Display the following *If...and...then...because* statement and ask the learners if they can discover the underlying code:

   If SCIENCE = 7243623 and INQUIRY = 4678479 *then*
   EXPLORE = __ __ __ __ *because* the 7288376 [__ __ _ __ __ __]
   is based on... (complete this sentence).

Optional Hint: What do all of these words have in common? What everyday device uses a seven-digit sequence to encode information?

b. Challenge the learners to use the code they discovered to determine which names of science disciplines and subfields can be represented by seven-digit telephone sequences.
c. Ask learners to suggest memory-related reasons for the following

attributes of telephone numbers: (1) Very early in the marketing of telephones (long before text messaging and cell phones), alphabet letters were overlaid on the numerical digits. (2) Telephone numbers are "chunked" (as are nine-digit social security numbers) and, within a given calling zone, consist of seven digits.

Experiment #3: Mnemonics Make Minutia Memorable

Learning science, like learning any discipline, requires becoming familiar with a number of common facts and terms. Scientists have invented acronyms (and acrostics) to describe and label new technologies and phenomena, such as light amplification by stimulated emission of radiation (LASER), radio detecting and ranging (RADAR), sound navigation and ranging (SONAR), self-contained underwater breathing apparatus (SCUBA), acquired immune deficiency syndrome (AIDS), and light emitting diodes (LEDs). Similarly, some common sequences in science can be remembered by creating mnemonics in the form of sentences. Here are some examples:

Order of Planets From the Sun:*

Mercury	Venus	Earth	Mars	Jupiter	Saturn	Uranus	Neptune	Pluto
My	Very	Easy	Method	Just	Set	Up	Nine	Planets

Biological Taxonomy:

Kingdom	Phylum	Class	Order	Family	Genus	Species
King	Philip	Came	Over	For	Great	Spaghetti

Although such mnemonics can be presented to students as examples, it is more important to teach students how to invent their own mnemonics. Invite learners to share other science mnemonics that they have devised or learned for the above sequences or for other science facts (see Internet Connections on memory).

Experiment #4: Meaningful Verbal Learning

(For comments and answers, see p. 138.)
For many students, trying to learn science is more difficult than

---

*This mnemonic may need to be modified because in late 2006 astronomers voted to downgrade the status of Pluto.

learning other "foreign" languages. Science instruction that is not grounded in fun phenomena first (with facts following) requires memorizing unknown "foreign" terms for unknown "foreign" things. Experienced science teachers have been immersed in the foreign land and culture of science for so long that they have forgotten the tyranny of terminology faced by students in science classes. The following activity helps teachers remember what it feels like to be a "stranger in a strange land."

Successively project each of the following vertical lists of nine terms on a whiteboard or screen, give learners 10–15 seconds to read and remember the list, cover the list, then ask them to write down as many of the terms as they can remember. Do not indicate that the sequence is important unless learners ask you and then indicate that they can remember the terms in any sequence they desire. Retention rates (correct spelling is required) should increase dramatically across the three lists even though the second and third lists contain more bits of information and would require more computer disc space to store. (The computer-brain analogy has severe limitations.)

WOT	MILK	MEANINGFUL
LOM	KITTEN	VERBAL
EAM	HOUSE	LEARNING
ZEL	JUMPED	IS
KAC	SCHOOL	AN
NIK	MANY	EASIER
BEN	STARS	TASK
GUR	WORK	THAN
FAM	SAIL	ROTE

When reviewing the results, ask: How can we explain the differences in our retention rates of these lists? What memory techniques did the best performers use with the first and second lists?

Debriefing

When Working With Teachers

The interactive nature of the teaching ←→ learning dynamic places responsibilities on both teachers and learners. The teacher, textbook, and other external resources serve as catalysts that work from the "outside in" to help learners in their quests for understanding. Learners must actively work from the "inside out" to expand and/or reconstruct their unique prior conceptions in light of new experiences and teacher feedback. A sports analogy might be useful here. Learning is not a spectator sport! Students-athletes have to do the heavy weightlifting work of learning. But teachers assist them by working as cognitive coaches who systematically control and scaffold, over time, the cognitive load demands of the curriculum. Teachers should also model for students how to learn more efficiently by intentionally making connections and searching for meaningful patterns. Teachers can learn more about memory enhancement techniques and research in books (see the excellent books listed in Extensions) and on the internet (Internet Connections: *American Educator*'s Ask a Cognitive Scientist; HowStuffWorks; the Wikipedia links).

The puzzles (Experiments #1–#4) also serve as exemplars of the nature of science. The Number Sequences and Patterns activity (Experiment #2) could be used to introduce the idea of genes as information codes and bioinfomatics (also see Activity #10 in this book).

Extensions

These books will help you explore more memory enhancement techniques: Bradley 2004; Buzan 1991; Higbee 2001; Lorayne 1990; Lorayne and Lucus 1996.

Internet Connections

- American Educator: (1) Ask a cognitive scientist column: *www. danielwillingham.com* (2) Allocating student study time: Massed versus distributed practice: *www.aft.org/pubs-reports/american_ educator/summer2002/askcognitivescientist.html* (3) Inflexible knowledge: The first step to expertise (or why transfer is hard): *www.aft.org/pubs-reports/american_educator/winter2002/CogSci. html* (4) Students remember…What they think about: *www.aft. org/pubs-reports/american_educator/summer2003/cogsci.html* (5) What Will Improve a Student's Memory?: *www.aft.org/ pubs-reports/american_educator/issues/winter08_09/index.htm*

• Exploratorium Online Exhibits: Don't forget—Playing games with memory: (1) *www.exploratorium.edu/memory/dont_forget/index.html* (2) Common cents: Can you identify the correct penny from a group of 11 other "imposters"?: *www.exploratorium.edu/exhibits/common_cents/index.html*

• HowStuffWorks: Human memory: *http://health.howstuffworks.com/human-memory.htm*

• Wikipedia: (1) Cognitive load theory: *http://en.wikipedia.org/wiki/Cognitive_load* (2) Cryptanalysis or code-breaking: *http://en.wikipedia.org/wiki/Cryptanalysis* (4) Ebbinghaus (memory researcher): *http://en.wikipedia.org/wiki/Hermann_Ebbinghaus* (5) Fibonacci sequence: *http://en.wikipedia.org/wiki/Fibonacci_number* (6) Gene expression: *http://en.wikipedia.org/wiki/Expressed_genes* (7) List science mnemonics: *http://en.wikipedia.org/wiki/List_of_mnemonics* (8) Memory: *http://en.wikipedia.org/wiki/Memory* (9) Mnemonics: *http://en.wikipedia.org/wiki/Mnemonics* (10) Spacing effect: *http://en.wikipedia.org/wiki/Spacing_effect*

Answers to Questions in Procedure, Experiments #1, #2, and #4

Experiment #1: Most learners will not be able to recall the entire 13-digit sequences for any of the examples *unless* they discover the underlying pattern and corresponding number "chunks":

1	**3**	**9**	**27**	**81**	**243**	**729**				Each successive number chunk is the previous number × 3.
1	**4**	**9**	**16**	**25**	**36**	**49**	**64**			Each successive number chunk is the square of the #1–8.
0	**1**	**1**	**2**	**3**	**5**	**8**	**13**	**21**	**34**	Each successive number is the sum of the two previous numbers; this pattern is called a Fibonacci sequence.

(*Note:* Once an underlying mathematical pattern is discovered, the 7 (+/− 2) limit is no longer a factor because the learner can extend the now meaningful pattern out to whatever length is desired.)

Experiment #2:

a. EXPLORE = 3975673 *because* the 7288376 [or PATTERN] is based on the fact that SCIENCE, INQUIRY, and EXPLORE are all seven-letter words as are telephone numbers within a given area code. In the two words SCIENCE and INQUIRY, there are seven unique single letter/number matches and one case of overlap (i.e., both Q and R are represented by the number 7). Learners will be unable to encode the word EXPLORE into its #sequence or decode the #sequence 7288376 into the word PATTERN unless they recognize that the pattern is based on the telephone code: 2: ABC 3: DEF 4: GHI 5: JKL 6: MNO 7: PQRS 8: TUV 9: WXYZ; #1 and the * and # symbol keys have no associated letters; 0 = Operator. On cellular phones, 1 translates to –; or @ and 0 can be + and the # key can be a blank space.

b. Other seven-letter science disciplines include BIOLOGY = 2465649, GEOLOGY = 4365649, and PHYSICS = 7497427. If three-digit area codes are used, ten-letter words can be "spelled" out, such as EXPERIMENT and HYPOTHESIS. Of course, students regularly use cell phone text messaging to spell out or abbreviate words of any length. The Federal Communications Commission is currently working on a nationwide emergency alert system that will use text messages delivered to cell phones.

c. When telephone numbers spell out words, they are easier to remember because they are more meaningful than a random number sequence. This is also why company ownership of internet web addresses that spell out the company's name is so important. Chunking numbers and limiting telephone numbers to 7 digits (= 3 + 4) (or 10 digits with area codes) reduces the memory demand. Some phone numbers also make memorable sequences of musical notes by eliciting our sense of rhyme and musical patterns.

[There are no questions or answers for Experiment #3.]

Experiment #4: The first column is most difficult in that, with the exception of the proper name Ben, none of the three-letter chunks are meaningful words in English. This is in contrast to the second list where each four- to six-letter chunk represents a familiar word (most are objects or actions that can be visualized) that can be strung together to create at least two meaningful sentences—for example, The KITTEN JUMPED for the MILK in the SCHOOL HOUSE; MANY STARS make SAILing WORK.

Ancient orators and today's memory experts encourage people to create meaningful and "fun" images and stories to help them remember, for example, people's names, the location of objects, or the sequence of different parts of a speech. The last list is obviously a no-brainer in that, by reading from the top to bottom, we discover that it is a single sentence that makes sense and therefore places a low cognitive load on short-term or working memory and is easily transferred into long-term memory.

Activity 13

Sound Tube Toys: The Importance of Varying Stimuli

Expected Outcome

A corrugated plastic Sound Tube is swung overhead at various speeds to produce its fundamental note (inversely proportional to tube length) and up to three resonant overtones (at faster speeds). The relationship between tube length and pitch (or frequency) is also explored.

Science Concepts

Sound is a form of energy created and transmitted as a vibration or mechanical wave that can vary in pitch (frequency) and volume (amplitude). Compression or longitudinal waves, such as sound waves, require a medium for transmission. For wind instruments, the longer the column of vibrating air, the lower the pitch. The activity described in the Procedure section can also be used to explore (1) fluid dynamics and Bernoulli's principle in physical science classes and (2) the evolutionary advantage of animal sensory systems geared to respond to novelty or changing circumstances (to detect danger or opportunity) in life science classes.

Science Education Concepts

Novelty and changing stimuli are means to activate student attention and catalyze cognitive processing. Simple toys can be used to engage interest and to develop and assess the skills of scientific inquiry and the basis of sound science concepts and "big ideas" (i.e., the Benchmarks for Science Literacy "Common Themes" [AAAS 1993] and the National Science Education Standards "Unifying Concepts" [NRC 1996]). Well-selected TOYS can lead to Terrific Observations and Yearnings for Science. Additionally, the particular toy in this activity can be used as a *visual* (and *auditory*) *participatory analogy* for why slow, monotone, noninteractive instruction is ineffective and conversely how planned variety can cause students to stop, look, listen, and question.

Materials

- 2 or more sound tubes from science education catalogs, science museums, nature stores, novelty shops, or dollar stores. Hardware and pool stores also sell *internally corrugated or ribbed* swimming pool hose that works nicely when cut into 2–6 ft. lengths. Noncorrugated hoses of equivalent lengths can be contrasted with the corrugated tubes. Tubes can be cut to smaller lengths (for higher pitches) or duct-taped together to make longer ones (for lower pitches). Other possible sources for sound tubes are the following:

- Arbor Scientific. *http://arborsci.co.* 800-367-66950. Sound Pipe. P7-7200. $2.75 (with instruction sheet). See also Arbor Scientific's Cool Stuff Newsletter archive for many related teaching ideas that are lots of fun: *www.arborsci.com/CoolStuff/cool14.htm#SingTube.*
- Educational Innovations. *www.teachersource.com.* 888-912-7474. #SS-600. $2.45 each.
- *Optional:* 1 chromatic, electronic tuner (the kind used for tuning guitars and other stringed instruments) if you want to measure the exact frequencies produced by different length tubes.

Points to Ponder

Learn to endure yourself to drudgery in science. Learn, compare, collect the facts! … Facts are the air of scientists. Without them you can never fly…. Without them your theories are vain surmises. Learning, experimenting, observing—try not to become the archivist of facts. Try to penetrate to the secret of their occurrence, persistently search for the laws that govern them.

—Ivan Pavlov, Russian physiologist (1849–1936)

General appeals to a child (or to a grown-up) to think, irrespective of the existence in his own experience of some difficulty that troubles him and disturbs his equilibrium, are as futile as advice to lift himself by his boot-straps.

—John Dewey, American philosopher-educator (1859–1952) in *How We Think* (1910)

Procedure

When working with teachers, set the stage for the activity by quickly eliciting ideas about key attributes of sound science teaching (as well as the science of sound). When working with teachers or students, pull out a Sound Tube and begin swinging it at a constant low speed to produce the tube's fundamental note. If time and available materials permit, distribute a number of the tubes for hands-on explorations in small groups. After

some time for free play, move the groups toward more systematic study of the phenomenon by challenging them to pose testable hypotheses (If… then…) about how the tube works. Different teams can be challenged to address different questions. Alternatively, the following questions and related investigations can be carried out via a teacher-directed, interactive participatory demonstration. In either case, ask the learners to predict-observe-explain the underlying principles of the science of sound.

(See answers to questions in steps #1–#4 on pp. 149–151.)

1. What forms of energy can you observe in this "system"?

2. How is the mechanical energy of your swinging arm changed or "converted" into sound? Think about what sound is and how this toy produces it. Possible inquiry questions include:

   a. Would a smooth, nonribbed, or noncorrugated tube (such as a section of garden hose) work to produce such a clear musical sound? Why or why not? Try it! If the smooth, noncorrugated tube is not the same length, diameter, or color as the corrugated tube, take the opportunity to discuss the idea of a fair test, relevant variables, and experimental controls.

   b. Do you think the sound is produced primarily by air moving through the inside and/or over the outside of the tube and why? How could you empirically test your hypothesis?

   c. What two ways could the air move through the inside of the tube? How could we determine if the air moves from the lower, almost stationary end to the higher, moving end or vice versa?

   d. How can the evidence gathered above help you develop an explanation for how this musical instrument works (i.e., what is its form-to-function relationship)?

3. What variations might we make in this instrument to vary its volume, pitch, and tonal qualities? What measurements might be useful to take as we make these variations?

   a. If we shorten the tube by cutting off some of the plastic, what will happen to the pitch of the instrument (constancy or change)? Conversely, how would a longer tube behave? (*Note:* A longer tube can be created by duct taping two or more sound tubes together end-to-end. How does this longer tube model stringed or wind instruments of varying length or scale?)

b. Will the sound change if the tube is rotated faster, and if so, how? A chromatic electronic tuner can be used to determine the actual notes produced. What other musical instruments can create higher pitch (or frequency) notes by having the musician blow harder or faster? (*Note:* The science of overtones could be explored in high school physics classes with oscilloscopes.)

c. What other experiments would you like to try with these toys?

4. Biology of Sound Sensing and Survival

a. Why do humans and other animals tend to tune out and stop noticing constant, unchanging background sounds? How can selective inattention be considered a useful adaptation? As you elicit ideas, begin to rotate the tube first faster and then slower to produce alternating higher and lower pitches.

b. How can these tubes be used to explore the biology of hearing in stereo? In terms of animal survival, why is out-of-sight, out-of-mind a maladaptive strategy in terms of evolution and natural selection? What competitive edge does good binaural hearing give to a predator (or prey)? How does a motion effect play out in terms of predator and prey adaptive behaviors?

Debriefing

When Working With Teachers

As an *auditory analogy,* consider how poorly designed and/or implemented science instruction is like the monotone pitch produced by a sound tube being rotated at a slow constant speed. Initiate teacher discussion with one or more of the following clusters of open-ended questions:

1. Is it the case, as Pavlov's quote suggests, that science learning has to contain a certain amount of drudgery? Should key facts be told to or discovered by students? How much of conventional schooling asks students to be "archivists of facts" versus "explorers and discoverers" of underlying principles? How should a science class look distinctly different from classes in other subjects?

2. During the course of a given lesson, a week of school, and a unit of instruction, how much novelty should be introduced to activate student attention and catalyze cognitive processing? What

roles can science toys play in a science classroom? Can there be too much levity and laughter in a science classroom? At some point does a science teacher risk becoming an MTV-type entertainer rather than an educator? How does a science teacher balance or integrate education with "edu-tainment"?

3. How can science teachers capitalize on the short-lived, reflexive attention generated by surprising discrepant events to motivate students to exert concentrated mental effort on the underlying scientific principles? Can breaking a 45–50 min. class into multiple segments with distinct, varied instructional strategies decrease the likelihood that students will tune out or become habituated to instruction? Conversely, how can we use the related phenomenon of acclimation to create a classroom environment in which positive learning behaviors are presented to students in a habitual manner because they are self-rewarding? (*Note:* The adaptation of selective attention helps explain why it is critical to maintain a relevant high signal-to-noise ratio in the classroom— it will gain and maintain student attention. It also points to the importance of teachers developing "acting skills" that involve intentional variation in such things as body movement, facial expressions, and vocalizations (see Tauber and Sargent Mester 2007 for an excellent introduction to "teaching theatrics"). Also, interested teachers can explore the research on student misconceptions about sound (e.g., see Driver et al. 1994; Chapter 18 is a good starting point).

When Working With Students

This activity can be used during either the Engage phase or the Elaborate phase of a 5E cycle (see Appendix B for more information on the 5E Teaching Cycle), either as a diagnostic assessment or formative assessment of basic concepts about sound energy. It can also be used in a unit on fluid dynamics and Bernoulli's principle (see Procedure, questions in steps #2c and #2d). Computer simulations (see Internet Connections: PhET) can be used to help students "see" sound waves and the effects of varying frequency, volume, and harmonic content.

Extensions

1. *A Straw Symphony: Oboe, It's Science!* Students can make their own homemade oboes by pressing down and snipping one end of a soda straw so that the straw tapers to a V-shaped pointed tip. If these "double reed" ends are flattened by being pulled between the students' teeth and then placed just inside their pursed lips and blown or buzzed, a steady musical pitch is produced. Higher notes can be reached by progressively snipping off the other end of the straw as the students continue to buzz the reed end. The shortest straws will produced the fastest vibration of air and the highest pitch (see Internet Connections: Whelmers #11). Double reed instruments include bassoons, English horns, and oboes; single reed instruments include clarinets and saxophones.

2. *Panpipes.* These instruments can be constructed with noncorrugated cardboard mailing tubes or PVC piping of varying lengths (see Internet Connections). Arbor Scientific, among other science suppliers, sells a set of rainbow colored, 8-tone scale Boomwhackers (P7-7400, $32) to explore open- and closed-pipe resonance.

3. *Metal Slinkies, Newton's Cradle* (suspended steel balls), *and Dominoes.* These toys can be used to demonstrate compression or longitudinal sound waves. Small, 1 in. diameter Slinkies work nicely on an overhead projector; standard-size Slinkies can be used on the floor or horizontally suspended from the ceiling. Newton's Cradle serves as a linear macroscopic model of the collision between molecules that transmit sound energy through a medium. Further, if a large number of dominoes are arranged radially outward from a central point and a ball is dropped in the middle, a 2D sound wave visual analog is produced. In Earth science, the fast, Primary, P-waves generated by earthquakes are, like sound, compression waves, whereas the slower, Secondary, S-waves are shear or transverse waves.

4. *Bernoulli Basics.* Many everyday phenomena (e.g., airfoils, spray/atomizer bottles, movement of shower curtains, operation of soda straws and vacuum sweepers) can be used to demonstrate Bernoulli's principle (see Internet Connections: Bernoulli's Principle and Physics Video Demonstrations).

Internet Connections

- Arbor Scientific's Cool Stuff Newsletter: *www.arborsci.com/ CoolStuff/Archives3.aspx*. (See Sound & waves and Pressure & fluids demonstration listings.)

- Bernoulli's Principle Animation: *http://home.earthlink. net/~mmc1919/venturi.html*

- HowStuffWorks: Vacuum cleaners: *http://home.howstuffworks. com/vacuum-cleaner.htm*

- PhET Interactive Simulations: Sound ("see" and adjust the frequency, volume, and harmonic content of sound waves): *http:// phet.colorado.edu/simulations/sims.php?sim=Sound*

- Phil Tulga's Music Through the Curriculum: Homemade instruments with math and measurement: *www.philtulga.com/ HomemadeMusic.html* (e.g., panpipes and xylophones)

- Physics Video Demonstrations on Bernoulli's Principle: *www. wfu.edu/physics/demolabs/demos/avimov/bychptr/chptr4_matter. htm#Bernoulli* and *http://faraday.physics.uiowa.edu/heat.html* (16 video clips on Bernoulli force)

- University of Virginia Physics Department: Sound activity stations: *http://galileo.phys.virginia.edu/outreach/8thGradeSOL/ SoundStationsFrm.htm*

- Whelmers #11: Straw oboes: *www.mcrel.org/whelmers/whelm11. asp* #12: Bernoulli cans: *www.mcrel.org/whelmers/whelm12.asp* #28: T.P. away (Bernoulli's principle): *www.mcrel.org/whelmers/ whelm28.asp*

- Wikipedia: (1) Bernoulli's principle: *http://en.wikipedia.org/wiki/ Bernoulli's_equation* (2) Habituation: *http://en.wikipedia.org/wiki/ Habituation* (3) Primary waves: *http://en.wikipedia.org/wiki/ P-wave* (Earth science/seismology)

Answers to Questions in Procedure, steps #1–#4

1. In motion, the sound tubes display mechanical or kinetic energy converted into sound.

2. The "how" of the energy conversion process requires an understanding of the nature of sound.

    a. A smooth, nonribbed or noncorrugated tube (such as a section of garden hose) will not produce this effect. Sound energy is produced by the mechanical vibration of matter and transmitted through a physical medium such as air. In this case, the vibration of air across the *inside* ribbing of the tube produces the fundamental tone of the tube.

    b. The sound is produced primarily by air moving through the inside rather than over the outside of the tube. This can be demonstrated if you tape or use your palm to cover either end of the tube. The majority of the sound stops despite continued rotation. Alternatively, use an old vacuum sweeper hose with prominent external but not much internal ribbing.

    c. In a manner analogous to a vacuum cleaner, small bits of torn-up paper or smoke from an extinguished candle will rise up into and through the tube from the lower, nearly stationary, handheld end. If air were moving down the tube, the paper would be blown away from the lower stationary end.

    d. Air is being forced over the internal, plastic ridges as it moves through the inside of the tube from the lower, nearly stationary, handheld end of the tube to the top, the freely rotating, lower pressure end. Bernoulli's principle would predict that because the free end of the twirling tube is moving faster than the end in your hand, it creates a lower pressure region relative to the nearly stationary end in your hand. The higher pressure air that is at the end in your hand forces air up into the tube. Be sure to avoid using the terms *suck* or *suction,* which create misconceptions. Instead, emphasize that air moves (*pushes* rather than *pulls*) from higher to lower pressure regions. If the speed is kept constant, air vibrating against the plastic ridges in the tube will emit a fairly constant pitch sound.

3. The tube length, speed of rotation, and frequency and volume of the sound produced could all be systemically altered, measured, and correlated.

   a. The fundamental frequency will be higher in the shorter tube. Specifically, if it is cut in half, it will sound an octave higher in pitch *if spun at the same speed*. Conversely, a tube will be an octave lower if the length is doubled. This same pattern is evident in the sequence of piccolo–C flute–bass flute and violin–viola–cello–stand-up bass.

   b. Higher speeds of rotation produce up to three other resonant overtones (e.g., the fifth, an octave higher, and another fifth). Some higher octave notes on flutes and brass instruments can be reached this way without the player altering his or her fingering.

   c. Other experiments include using a hair dryer to blow air through the tube and noting the effects of high versus low air flow rates and hot versus room temperature setting. Hold the hair dryer an inch or two back from the end of the tube.

4. *Biology of Sound Sensing and Survival*

   a. From an evolutionary, cost/benefit point of view, being able to detect changing environmental stimuli to identify friend, foe, food, or fun (e.g., sex) is a higher priority than continuing to notice constant, nonchanging, nonmoving, noninviting, or nonthreatening sounds. Animals tune out (or becoming habituated to) such sounds as being not relevant from a survival (and reproduction) perspective (i.e., animals notice on a need-to-know basis). Similarly, the vision of many predators is much more attuned to moving than stationary objects (which explains the "freezing" and protective coloration adaptations of animals who are their prey). Mental energy and physical activity are conserved for things that are more likely to be important to survival and reproduction. The vast majority of sensory information that the brain receives is not consciously noted or cognitively processed and therefore rapidly decays with minimal lasting neurological traces. Selective inattention is in fact, an adaptive feature of our brains, which otherwise could be swamped in processing nonrelevant information.

b. Stereo hearing can be explored by having the learners shut their eyes, after which you move to alternative locations in the room and ask them to point in the direction of your location. Alternatively, suspend two linked-sound tubes with a string from a stand so that a wide end can be placed against each ear of a test subject. If that person closes his or her eyes and the tube is tapped off-center, the ears can detect the small difference in travel time and determine which side was tapped. Hearing in stereo, like stereoscopic vision, has significant evolutionary advantages for locating possible predators, prey, or mates in a 3D landscape that may offer visual hiding places. Echolocation (as found in bats and dolphins) takes this good idea one step further.

Activity 14

Convection:
Conceptual Change Teaching

Expected Outcome

Two bottles of different-colored water are placed vertically, one on top of the other with open ends together. They remain stable if the bottle with the hotter (red-colored) water is placed on top of the bottle with the cooler (blue-colored) water. If they are arranged in the reverse order, mixing occurs rapidly to form the color purple, which becomes uniformly distributed within the two bottles.

Activity 14

Safety Notes

1. Students and teacher should wear indirectly vented chemical splash goggles during this activity.
2. Sharps hazard: Glass thermometers are fragile and can shatter.

Safety Note

Use caution to avoid getting burned with hot water.

Science Concepts

Thermal-induced density differences cause "heavier" cooler fluids to fall and push warmer "lighter" fluids up, creating circulation in convection cells. Real-world applications occur in the atmosphere (e.g., both normal weather patterns and thermal inversions), the hydrosphere (e.g., seasonal turnover in lakes), and the lithosphere (e.g., movement of magma and plates).

Science Education Concepts

Prior conceptions always form the backdrop and foundation for new understanding. This *visual participatory analogy* suggests that when students' prior conceptions are activated and assessed, they are often found to be mixed up or scientifically inaccurate. Such misconceptions need to be overturned in light of carefully designed explorations and logical arguments. Compare/contrast discrepant-event activities introduce novelty and changing stimuli that lead students to skeptically review the validity of their prior conceptions.

Materials

- 4 identical glass bottles with ~3 cm mouth diameters (e.g., 12 oz. clear, colorless juice bottles or 250 mL Erlenmeyer flasks)
- Red and blue food coloring
- 2 credit card–size plastic cards or glass plates
- Thermometer (non-mercury)
- White backdrop

The speed of mixing in the convection cell is proportional to the temperature difference. Hot and cold tap water may be sufficient, but otherwise use ice cubes to cool down the tap water; use a burner or microwave to heat it up before the learners see the setup. Don't make the water too hot to touch.

- *Alternatively:* Use 2 water-filled, 1 L plastic soda bottles that can be connected by a 2-hole Fountain Connection (sold by Educational Innovations. #SS-10. $3.75).
- *Optional Extension Activity* (on p. 158): Hot Air Balloon. Use a large, clear, colorless, plastic garbage bag; several straws; a heat gun; a mylar helium balloon; and a hair dryer.

Points to Ponder

My business is to teach my aspirations to conform themselves to fact, not to try and make facts harmonize with my aspirations…. Sit down before fact as a little child, be prepared to give up every preconceived notion, follow humbly wherever and whatever abysses nature leads, or you will learn nothing.

—Thomas Henry Huxley, English evolutionary biologist (1825–1895)

You can't turn a thing upside down if there's no theory about it being the right way up.

—G. K. Chesterton, English author (1874–1936)

Procedure

(See answers to questions in steps #2–#6 on pp. 160–161.)

1. Before the learners enter the room, place several drops of red food coloring in two of the bottles and blue food coloring in the other two bottles. Slightly overfill the two bottles with red food coloring with hot water and the two bottles with blue food coloring with cold water (to create a convex meniscus). Slide a plastic card over the top of the two bottles that are to be inverted; invert one bottle with the red, hot water over one bottle with the blue, cool water and one bottle with the blue, cool blue water over the other bottle with the red, hot water. Leave the plastic cards in place. Place the 2 two-bottle setups in front of a white backdrop for optimal visibility. Alternatively, if the two-hole Fountain Connections and plastic soda bottles are used, plastic card separators are not needed, but you will need to perform the connection and inversion "live" after the students arrive and class begins. (*Note:* Standard, one-hole Tornado Tube connectors will not work as well as the two-hole connectors).

2. Ask the learners to use a think-write-pair-share to *predict* what will happen when you carefully remove the plastic cards while holding the top bottles in place. Also challenge them to indicate why they believe this will occur (i.e., use logical arguments that include an explanatory mechanism). Record their ideas, but resist answering their questions about the nature of the fluids other than to indicate that the same-colored fluids are exactly the same fluid in both setups.

3. Remove both of the plastic cards and ask the learners to *observe* the two systems for signs of constancy and change.

4. Have learners work in teams of three or four to *explain* these results with at least two alternative hypotheses about the nature of the fluids that would account for the different results with the two setups. If the observed results run counter to their predictions, challenge them to discover the flaw in their prior understandings. If their prediction was correct, does this mean that their explanations were necessarily sound? What scientific equipment and measurements would be needed to decide which hypothesis leads them to an empirically verified correct explanation?

5. Disassemble the two setups (over a sink or by reinserting the plastic card and lifting the top bottle off) and use a thermometer to check the temperature of the water in the four bottles (it will be uniform in the mixed-up, purple bottles and different in the unmixed bottles). Pose these questions: We say that "heat rises." Why is it more appropriate to say that cold fluids are denser than warm fluids and therefore "fall" and displace warm fluids upward? Why does a given volume of cold water contain more mass than the same volume of warmer water? Would this convection phenomenon "work" in a zero or microgravity environment such as an orbiting space shuttle?

6. Brainstorm a list of scientific concepts and real-world applications that are related to this demonstration.

Debriefing

When Working With Teachers

Discuss the role of novelty and discrepant events in activating attention and eliciting prior knowledge. The latter is a necessary diagnostic step to uncover what foundational concepts can be built on and what prior conceptions need to be challenged and partially or wholly "overturned." Use the two quotes to discuss the interactive relationship between experimental observations and theories; consider how theories are informed and changed by experimental observations. Note that the human senses are biologically adapted to notice changing stimuli (to avoid threats and to act on opportunities) and to turn off attention to nonchanging environmental stimuli. This evolutionary reality is a key principle used by advertisers, movie producers, MTV, and effective teachers! Starting instruction with engaging phenomena that create a need-to-know is a powerful means to generate the emotional buy-in and cognitive hook that "sells science." Discrepant events get students to move from reflexively paying attention to external stimuli to intentionally focusing on underlying science concepts.

The idea of comparing two setups that differ in only one respect and then asking learners to go through one or more predict-observe-explain cycles is a classic experimental design that can be used to teach many science concepts. Also, interested teachers can explore the research on (1) student misconceptions about heat-energy and density-floating/sinking (e.g., see Driver et al. 1994, chapters 19 and 20 and chapters 8 and 12, as starting points) and (2) the related, but broader based, conceptual change teaching theory (see Internet Connections).

When Working With Students

This activity could be used either for diagnostic (Engage-phase) or formative (Elaborate-phase) assessment of students' understanding of the topics of fluid density and/or thermal-driven convection cells. If this activity is used during the Engage phase, do not rush to prematurely provide Explain-phase "answers" without more Explore-phase activities. (For an explanation of the 5E Teaching Cycle, see Appendix B.) The activity also can be used in a meteorology unit

on weather and the ground-level concentration of air pollutants in thermal inversions or in units on hydrology (limnology) or geology (plate tectonics).

Extensions

1. *Homemade Hot-Air Balloon.* Tape some straws around the open end of a large clear plastic school garbage bag. The straws act as stabilizers. Enlist the aid of several volunteers (to hold the top and perimeter of the bag) and fill the bag from the lower open end with hot air using a paint stripper–type heat gun. A hair dryer may also work if it has a high heat–low volume setting and you use a lighter weight dry cleaning bag. Once the bag is expanded to full volume with hot air, turn the heat gun off before releasing the bag (or intentionally do not do this the first time to see if students will misconstrue that the air stream, rather than density, is the primary driving force). The bag will rise to the ceiling, move around with room air currents, and remain aloft for several minutes (especially in a room with a high ceiling). Hot-air balloonists know to use either earlier morning (e.g., 6 a.m.) or early evening (e.g., 6 p.m.) launch times to get the maximum temperature differential with minimum use of propane fuel to heat the air in their balloons.

   As a variation, you can "revive" a sagging, sealed helium balloon by using a hair dryer to increase its volume and decrease its density (see Internet Connections: Becker, Exploratorium, PhET simulation, UVA: Air Density, and Whelmers sites). A verbal analogy can be drawn for teachers: Effective science instruction "raises" questions, "expands" interest and motivation, and "lifts up" sagging attention and effort in learners.

2. *Lava Lamps* also work by convection. See Internet Connections. Also see Boomfield (2007), a useful reference for explaining a large number of real-world physics applications, as are the HowStuffWorks websites (see Internet Connections).

Safety Notes

1. Heat sources can burn the skin. Also, keep heat sources away from flammable materials and combustible materials.

2. Keep indoor hot-air balloons clear of fire sprinkler heads on the ceiling.

Internet Connections

- Becker Demonstrations: Instant hot air balloon (heat a partially deflated helium-filled, mylar balloon; it enlarges and rises): *http://chemmovies.unl.edu/chemistry/beckerdemos/BD059.html*

- Conceptual Change Teaching (overviews of theory and research from various sources): (1) *www.physics.ohio-state.edu/~jossem/ICPE/C5.html* (2) *www.ericdigests.org/2004-3/change.html* (3) *http://narst.org/publications/research/concept.cfm* (4) *http://projects.coe.uga.edu/epltt/index.php?title=Conceptual_Change*

- Doing Chemistry (movies of chemical demonstrations): Density of warm and cool water: *http://chemmovies.unl.edu/chemistry/dochem/DoChem004.html*

- Exploratorium Museum Snackbook: (1) Convection current: Make your own aquarium heat waves: *www.exploratorium.edu/snacks/convection_currents* (2) Hot air balloons: *www.exploratorium.edu/lc/balloons/index.html* (3) Pie pan convection: *www.exploratorium.edu/snacks/pie_pan_convection/index.html*

- HowStuffWorks: Liquid motion lamps: *http://home.howstuffworks.com/lava-lamp.htm* and Hot air balloons: *http://science.howstuffworks.com/hot-air-balloon.htm*

- Hot Air Balloons: Links, history, and models: *www.kathimitchell.com/balloons.htm*

- NARST Research Matters to the Science Teacher: *http://narst.org/publications/research.cfm*

- NASA: Candle flame in microgravity (classroom demonstration/experiment): *http://quest.nasa.gov/space/teachers/microgravity/9flame.html*

- PhET Interactive Simulations: Balloons (hot air and helium) and buoyancy (and gas laws): *http://phet.colorado.edu/simulations/sims.php?sim=Balloons_and_Buoyancy*

- University of Iowa Physics and Astronomy Lecture Demonstrations (video clips): Heat and fluids*: http://faraday.physics.uiowa.edu* (five demonstrations on convection)

- University of Virginia Physics Department: Convection currents, density and temperature (hands-on-experiences): *http://galileo. phys.virginia.edu/outreach/8thGradeSOL/ConvectionCurrentsFrm. htm* and *http://galileo.phys.virginia.edu/outreach/8thGradeSOL/ AirDensityFrm.htm*
- Whelmers #33 Density Balloon: *www.mcrel.org/whelmers/ whelm33.asp*
- Wikipedia: (1) Convection: *http://en.wikipedia.org/wiki/Convection* (2) Lava lamps: *http://en.wikipedia.org/wiki/Lava_lamps* (3) Misconceptions about the brain: *http://en.wikipedia.org/wiki/List_of_ misconceptions_about_the_brain*

Answers to Questions in Procedure, steps #2–#6

2. Many learners will predict that the red- and blue-colored fluids will mix in both setups to form the color purple. This prediction is likely because of learners' prior experiences with physically mixing different-colored paints or food coloring that are at the same temperature (i.e., red + blue = purple) in elementary art classes. Learners are less likely to have had many previous experiences with liquids that are either immiscible (e.g., oil and water) or miscible but can be layered due to density differences (related either to differences in temperature or to the concentration of a dissolved solute).

3. When the plastic cards are removed, the fluids will instantly mix in the thermally inverted setup (i.e., the one with the hot water on the bottom) and as thermal equilibrium is reached, the system assumes a uniform purple color and a uniform temperature. The noninverted, thermally stable setup will remain as two separate fluids with the red, lower density, hot water sitting on top of the cool blue water for longer than a class period. This latter effect occurs because diffusion-driven mixing is quite slow in layered liquids with a stable density gradient and because heat transfer via conduction is very inefficient and slow in liquids.

4. Possible hypotheses include the following: the two fluids are two miscible liquids with different densities such as alcohol and water; they are both water, but one is pure water and the other is saturated with dissolved salt or sugar; and they are both water, but at different temperatures. If time permits, learners can be challenged to complete another predict-observe-explain cycle to test how these other, non-heat-driven density gradients would turn over (or not) depending on the location of the more dense fluid. In any case, measurements of the red-blue-purple fluids can be taken with a balance and graduated cylinder to determine density and a thermometer to determine temperature,

5. Given the closer molecular packing, cold fluids actually contain more molecules (and therefore mass) per equal volume than warm fluids (if the same chemical substances are compared). In an environment that includes gravity, this thermally induced density gradient will create a convection cell and circulation when the more dense fluid is on top. In the absence of gravity, any system that depends on convection-driven movement (e.g., burning candles) will not work in the same manner.

6. Convection cells are major driving forces within the "solid" Earth and in its waters and atmosphere and are therefore a major explanatory factor in the Earth sciences. They also are important in the design of home heating, seasonal ceiling fans, and air-conditioning systems. Science supply companies sell convection-ventilation boxes and convection of liquids tubes for demonstrations.

Brain-Powered Lightbulb: Knowledge Transmission?

Expected Outcome

What appears to be a standard, ordinary lightbulb screwed into a portable flood-light or spotlight socket lights up when it is "plugged" into the presenter's ear! If unscrewed, it lights up when held in the presenter's hand, but will not do so for other people. (*Safety Note:* NEVER connect this battery-powered "magic" lightbulb into an actual live wall socket. The person holding it might receive a serious or fatal electric shock. When the demonstration is over, store the "magic" lightbulb in a place that students do not have access to.)

Science Concepts

This activity explores the closed, battery-powered circuits that convert chemical potential energy to electrical energy to light energy. Teachers can also have students discuss the importance of skeptical review in science. In biology classes, the activity can be used to playfully introduce the topic of artificial photosynthesis as a potential future source of energy.

Science Education Concepts

Since Edison's invention of the lightbulb in 1879, it has been commonly used as a visual symbol of intelligence, ingenuity, and "bright ideas," expanding on an ancient association of light with knowledge and understanding. As a *visual participatory analogy*, the message is that the teacher's "light" cannot simply be given to students but rather serves as a spark to activate students' attention and catalyze cognitive processing. Puzzles and counterintuitive or discrepant-event events are powerful activities because they promote emotional curiosity and cognitive reconstruction efforts when guided by appropriate questions.

Materials

The "magic lightbulb" was popularized by Uncle Fester in the 1960s *Addams Family* TV series (and in late-1990s movies). The 1967 version was made of plastic, powered by a replaceable 1.5 V battery, and would *not* fit into a standard socket (for safety reasons).

Today the bulbs are available from various toy, novelty, and magic stores in smaller, plastic versions and realistic glass, standard-size versions. The realistic glass version (along with other science-based magic tricks) may be purchased over the internet from various suppliers:

- Abracadabra Magic. *www.abra4magic.com.* 732-805-0200. A079. $9.99
- Creative Presentation Resources, Inc. Magic Supplies: Atomic light bulbs #103-026. $5.95. *www.presentationresources.net/magic_supplies. html.* (See Light Fantastic variations.)
- GagWorks (magic lightbulb): *www.gagworks.com.* #GW1158. $8.99
- Magic Supply Company. *www.magicsupply.com.* 816-224-5000. FL120. $12.99
- Magic Trick Store. *www.magictrickstore.net/magiclightbulb.* $9.95

A small piece of aluminum foil, metal ring, or steel wool secretly held in the hand can be used to complete the circuit and turn the bulb on and off at will. (Steel wool should be handled only by the teacher, not by the students.) A socket with a short section of wire and plug (such as from a portable spotlight or floodlight) can be used as a fake circuit to "plug into the teacher's brain" via his or her ear.

Points to Ponder

A wise man, therefore, proportions his belief to the evidence.

—David Hume, Scottish philosopher (1711–1776) in *An Enquiry Concerning Human Understanding* (1748)

With regard to the electric light, much has been said for and against it, but I think I may say without fear of contradiction that when the Paris Exhibition closes, electric light will close with it, and no more will be heard of it.

—Erasmus Wilson, British physician and Oxford University professor (1808–1884)

Light is always propagated in empty space with a definite velocity, "c," which is independent of the state of motion of the emitting body.

—Albert Einstein, German-American physicist (1879–1955)

Procedure

The difference between performances by magicians and magical and memorable science teaching is that the teacher intentionally scaffolds a dialogue of discovery (which may extend over a series of lessons in a 5E Teaching Cycle; see Appendix B) to lead students to discover the underlying scientific explanation behind the "magic trick."

Like most magic tricks, this one works best with some initial, intentional misdirection. Introduce the activity with language such as the following: "The relationship between the wisdom of the teacher and student learning is not as simple or direct as commonly assumed. Let's consider the lightbulb as a symbol of teacher intelligence. As you can see, I can screw this lightbulb into a socket from a flood light and plug it into my superior brain and it will light up." (*Performance Note:* Screw the bulb into the socket at the same time you bring the plug end up to your ear. Then, slightly loosen the bulb as you let the plug fall away from your ear. Alternatively, holding the bulb in your cupped hand that contains a hidden piece of steel wool, say, "I can also focus my mental energy on the bulb alone and make it light up.")

After this "magic routine," give a similar looking, but standard lightbulb to a volunteer (or use the magic one without the piece of steel wool); he or she, of course, will not be able to light the bulb by sticking it in his or her ear. (**Safety Note:** Do NOT allow students to handle the socket–magic lightbulb setup, even though they will want to. The risk is too great that they will plug it into a wall socket.)

When Working With Teachers

Play off this visual participatory analogy by asking open-ended questions such as the following:

1. Does "knowledge propagation" from the teacher to the "empty space" of students' brains happen independently of students' relative motion or actions (see Einstein's quote)?
2. Can knowledge be transmitted from a teacher and received by a student without direct mental activity of the student?
3. Can students' brains really be considered empty spaces?
4. What's wrong with the ideas of knowledge transmission and reception as a model for teaching (as telling or preaching) and learning (as listening)?

5. You may wish to use the quote about the future of the lightbulb to point out that even highly educated scholars believe things that are later shown to be not the case. In a similar vein, teachers need to remain open to future research and new technologies that may challenge traditional pedagogical practices and ideas about learning.

When Working With Students

With a modified introductory patter (e.g., "Today we are going to begin to explore a mysterious source of energy that pervades the universe and also resides within us…"), this participatory demonstration can be used as either an Engage or Elaborate activity in a unit on electrical circuits. Ask questions such as the following:
(For answers to questions in steps #1 and #2, see p. 170.)

1. Does the human brain depend on the generation, transmission, and use of electrical current? Is the electrical current generated within the human brain large enough to power a standard incandescent lightbulb? If not, can it be detected and measured by other means? What medical benefits might be associated with these kinds of technologies?
2. What are the two common sources of electrical power for lighting bulbs? Which of these is most likely used in this activity? How might you account for the inability of your classmate to replicate the teacher's demonstration?
3. Alternatively, challenge students to work in small groups to (a) draw what they think the inside of the lightbulb looks like and/or (b) come up with their own list of 20 questions that you will only answer with yes or no.

Debriefing

When Working With Teachers

Discuss the open-ended, science-inquiry questions above. Remember that using puzzles and discrepant-event phenomena to motivate and activate student attention and catalyze cognitive processing is critical to good instruction. The quote by Hume can be used to point to the importance of logical argument and skeptical review in science. In

learning, the challenge is the need not just to fill in the conceptual holes of what we don't know but also to activate and confront what we know that isn't so. Internet Connections has several good resources about cognitive neuroscience and misconceptions about the brain.

When Working With Students

Teacher questioning and "closure" needs to be handled differently depending on where in a 5E cycle the activity is used. If the activity is used in the Engage phase, it is important to intentionally avoid premature closure. Instead, let students do Explore- and Explain-phase activities before returning to solve the mystery of the brain-powered lightbulb. Activity #9 in this book makes for a good follow-up, Explore-phase activity.

Extensions

1. *Bio-Energy: A Bright Idea*. Hold up a green plant and discuss how green plants use photosynthesis to capture and convert solar light energy into chemical potential energy (sugar). Raise the question: Can green plants also produce electricity? Use the same lightbulb screwed into a spot or flood lightbulb socket with the plug inserted into either a leaf of a green plant or a beaker of green-colored water (to simulate a solution of chlorophyll). Simultaneously screw the bulb into the socket and touch the leaf (or green water) and the lightbulb will light. Slightly unscrew the bulb as you pull the plug away from the leaf or green water (see Internet Connections: Wikipedia: Artificial Photosynthesis). Of course, biofuels can be burned to generate electric power, and fossil fuels are, in fact, biomass that was chemically converted over millions of years but still contains much of its original "trapped solar energy."

2. *Alternative Brain-Power-Lightbulb Activity*. Although the standard-size Magic Light Bulb is powered by a hidden battery, it is possible to use the "brain power" (or more accurately, static electricity generated by rubbing a balloon against human hair to produce over 10,000 volts of EMF) to light up a smaller, "special" bulb that has very low current (amperage) and power (wattage) demands. Edmund Scientific sells a Static Electricity Bulb/Human

Powered Light, #3081446, $5.95 (*http://scientificsonline.com*). A "brain-powered," electrically charged balloon also very briefly lights up a disconnected fluorescent tube held in your hand in a darkened room.

3. *Cognition and Creativity*. The website *www.optillusions.com/ dp/1-19.htm* has a black/white reverse image (or "negative after-image") of a lightbulb. If you stare at it for 30–60 seconds and then shift your gaze to a white wall or screen, the image will reappear as a glowing lightbulb! Use this discrepant event as an entrée into a discussion of human visual perception as related to cognition. Alternatively, discuss the role of human creativity in designing more energy-efficient lighting (e.g., compact fluorescent bulbs and LEDs).

Internet Connections

- Arbor Scientific's Cool Stuff Newsletter: *www.arborsci.com/CoolStuff/ Archives3.aspx*. (See Electricity and Electrostatics demonstrations.)

- International Mind Brain and Education Society: *www.imbes.org*

- Society for Neuroscience: Brain Facts: A primer on the brain and nervous system: 74-page book, CD, and free pdf file: *www.sfn.org/ index.aspx?pagename=brainfacts*

- Twenty Brain Buster Q&A on Electrical Circuits (for grades 12-and-up physics instruction): *http://courses.science.fau. edu/~rjordan/busters_26/push-ups_3b.htm*

- The Promise of Artificial Photosynthesis (article): *www.energy bulletin.net/317.html*

- Wikipedia: (1) Artificial photosynthesis: *http://en.wikipedia.org/ wiki/Artificial_photosynthesisv* (2) Cognitive neuroscience: *http:// en.wikipedia.org/wiki/Cognitive_neuroscience* (3) Misconceptions about the brain: *http://en.wikipedia.org/wiki/List_of_misconceptions_ about_the_brain*

Answers to Questions in Procedure (When Working With Students), steps #1 and #2

1. Yes, the neural networks in the brain are part of a complex electrochemical system that generates and responds to electrical current. The current produced at any one time in any specific location is thousands of times too small to light a standard incandescent lightbulb. However, electroencephalography (EEG) and magnetoencephalography (MEG) can directly monitor the ever-changing electrical activity in the brain. Newer technologies—such as positron emission tomography (PET) and functional magnetic resonance imaging (fMRI)—indirectly measure neural activity based on the fact that when neurons fire they use oxygen and glucose at a greater rate, which in turn stimulates increased blood flow to the active areas of the brain. All of these technologies are contributing to improved medical diagnoses and treatments for brain-related problems, as well as to neuroscience research. As a body organ, the brain consumes energy (using oxygen to burn glucose) at a rate that is disproportionate to its volume and mass. That is, a 3 lb. brain that equals 2% of a 150 lb. person will use approximately 25% of the body's glucose and oxygen. But, it still requires less power than a 25 W lightbulb!

2. Alternating current from a wall socket or direct current from an electrochemical cell (battery), solar cell, or handheld generator is used to power lightbulbs. A hidden DC battery is located inside the Magic Light Bulb. A lightbulb will light only if it is part of a complete, closed circuit. The teacher is using some hidden piece of metal to complete the circuit or has the lightbulb completely screwed into the socket so both contact points on the bulb are connected in the closed circuit.

Activity 16

Air Mass Matters: Creating a Need-to-Know

Expected Outcome

A flat pinewood stick (or other soft wood) is placed under several sheets of newspaper and extended over the edge of a table. It snaps when quickly struck, without lifting or tearing the paper.

Science Concepts

Air has weight and exerts a pressure of 10 N/cm² (or 14.7 lbs/in²) at sea level. Gases are not "no thing." Gases have mass, occupy space, exert pressure, and are composed of molecules separated by truly "empty" space. Inertia, or the tendency of a body at rest to stay at rest, is also a relevant factor in this experiment.

Science Education Concepts

Teachers sometimes need to initially take familiar (and therefore unnoticed) things and make them strange so that they can become familiar again but—the second time around—understandable. Discrepant or counterintuitive events activate learners' attention and catalyze cognitive processing by creating need-to-know motivation. This demonstration serves as a *visual participatory analogy* in the sense that students/sticks can only successfully lift the conceptual weight or load of a given educational task if the instructional pace (or speed) that they are expected to move at is within their zone of proximal development or ZPD (i.e., what the learner can achieve based on prior knowledge and abilities with the scaffolding provided by a carefully targeted instructional sequence and a supportive teacher). In the case of this demonstration, if the teacher pushes the student/stick at a too fast a rate, it breaks. If the teacher wants to avoid breaking the student/stick, he or she needs to use a slow, deliberative pace rather than a forceful, quick pass through too many topics in too little time (see Internet Connections: Wikipedia: Cognitive load theory and ZPD).

Materials

- Flat pinewood stick (e.g., cheap yardstick or extra long [2 ft.] paint stick) and several sheets of newspaper

Safety Note

Students and teacher should wear safety glasses or goggles during this activity.

Points to Ponder

I do not mind if you think slowly. But I do object when you publish more quickly than you think.

—Wolfgang Pauli, German-American physicist (1900–1958)

When you believe you have found an important scientific fact, and are feverishly curious to publish it, constrain yourself for days, weeks, years sometimes, fight yourself, try and ruin your experiments, and only proclaim your discovery after having exhausted all contrary hypotheses.

—Louis Pasteur, French chemist and microbiologist (1822–1895)

Procedure

(See answers to questions in steps #1–#4 on p. 177.)

1. Place the pinewood stick on a table with about 10 cm (4 in.) extending over the edge. Ask: What would happen if I were to strike the extended end of the wood? Do this experiment.

2. Repeat the experiment, except this time place two, full sheets of standard-size newspaper on top of the portion of the wood stick that rests on the tabletop, taking care to smooth out the newspaper and press it down firmly against the tabletop. If this is not done and a significant air pocket resides under the paper, the demonstration will not work consistently as intended due to an equalization of air pressure above and below the paper. Again, ask the learner to *predict* what will happen when you rapidly strike the extended portion of the stick.

3. Ask questions such as the following:

   a. What did you *observe* in this second case and how can you *explain* the difference between these two trials?

   b. What would happen if I placed the wood stick on the table without the newspaper and had someone press down on the portion that rests on the table while I strike the extended portion?

   c. What would happen if I used the newspaper again, but rather than striking the stick, I slowly pressed down on it?

   d. How do these extra tests provide clues as to how to explain the demonstration in which the stick breaks?

   Any of these variations can be repeated with new sticks (or by extending the stick if it is long enough).

4. Depending on where this activity is used in a 5E science unit (i.e., Engage versus Explain or Elaborate; see Appendix B for a discussion of the 5E Teaching Cycle) and the grade level, the teacher may have students calculate the effective weight of air that is pressing down on the surface area of a single piece of newspaper.

Debriefing

When Working With Teachers

In a discrepant-event demonstration, the teacher takes something that is typically unnoticed by students (e.g., air pressure) and makes them pause, perceive anew, and ponder on this thing (i.e., it activates their attention). Discuss the pedagogical advantages of using the phenomena-before-facts or the wow-and-wonder-before-words approach over the common (reverse) approach in which the teacher starts with lecture notes or gives a reading assignment on air pressure. Teachers can explore the large body of published research on student misconceptions about gases, air, and pressure (e.g., see Driver et al. 1994; see chapters 9 and 13 for overviews).

The demonstration serves as a visual participatory analogy for how the cognitive load of a given educational task or learning objective—as perceived by students—depends, in part, on the speed of instruction. Rushing through big ideas too quickly can "break" students, whereas a slower, more deliberate approach is much more likely to succeed. The contrasting quotes on page 173 can focus learners' attention on the nature of science (i.e., empirical evidence, logical argument, and skeptical review) and can help them contrast the relative checks on truth in the popular press versus such checks on scientific journals.

The best science teaching is more about inspiring inquiry than indoctrination in "received truth." Similarly, if students are going to be asked to calculate the weight of air pressing down on the paper, it is important that the teacher first create a context that catalyzes learners' curiosity—rather than present calculations in no context at all, an approach that will kill curiosity.

When Working With Students

After students do other related activities—such as the Extensions and activities found on websites listed in Internet Connections—that make the unnoticed effect of air pressure "sensible," the teacher should introduce the basic facts about air pressure discussed in the Answers to Questions in Procedure, step #3 on p. 177.)

Extensions

1. *The Crushing Soda "Pop" Can.* This discrepant event is a variation of an old demonstration. A little water placed in an empty 1 gal. rectangular metal can is brought to a boil and the can is then removed from the heat and tightly capped. As the can cools, the water vapor condenses and leaves a partial vacuum inside the sealed can that then collapses under the now greater, external atmospheric pressure. With a soda can, just cover the bottom of the empty can with water, boil it to drive out air, and fill the can with water vapor. Then either cap the can with a fizz-saver lid or turn the can upside-down on top of a container of water. In the latter case, the can will rapidly crush and partially fill with water. (See Internet Connections: Purdue University, among other websites, for explanations.)

   Alternatively, a vacuum pump causes the reverse expansion effect by decreasing external pressure on a partially sealed, air-filled container (e.g., a balloon, a marshmallow, or shaving cream) that is under an evacuated, airtight chamber.

Safety Note

The edges of metal cans can be sharp and can cut the skin. Handle with caution.

2. *Air Mass Matters.* Place two identical, uninflated balloons on a double-pan balance. (See safety issues regarding latex balloons on p. 89. Avoid latex balloons.) They will balance. If one of the balloons is then inflated and tied off, the balance will tip in direction of that balloon, indicating that air has mass. Similarly, a teacher can demonstrate that if a deflated sports ball is weighed and subsequently pumped up with air, the mass gain is directly proportional to the number of pumps. Alternatively, this can be done on a smaller scale as a hands-on exploration by using fizz-saver caps and a 2 L empty soda bottle. (*Note:* P1-2050/Individual Pressure Pumper can be purchased from Arbor Scientific for $3.25 or from local stores as a device to save the fizz on opened soda bottles). In either case, you may want to use temperature strips on the plastic bottle to also study the relationship between pressure and temperature.

Internet Connections

- Arbor Scientific's Cool Stuff Newsletter: *www.arborsci.com/CoolStuff/Archives3.aspx.* (See Chemistry: Gas laws smorgasborg and Pressure and fluids demonstrations.)

- Can Crush Demo/Railroad Tank Car Crush: *www.delta.edu/slime/cancrush.html*

- HyperPhysics, Department of Physics and Astronomy, Georgia State University: Select Video/Demos: Fluids: Liquids and gases: Atmospheric pressure: *http://hyperphysics.phy-astr.gsu.edu/hbase/hframe.html*

- Purdue University: Can crusher: *http://chemed.chem.purdue.edu/genchem/demosheets/4.8.html*

- University of Iowa Physics and Astronomy Lecture Demos. *http://faraday.physics.uiowa.edu* (See Heat and fluid: Atmospheric pressure demonstrations: Crush the can, crush the soda can; Magdeburg hemispheres; Water column-water barometer; Suction cups-rubber sheets; Stick and newspaper and the vacuum cannon.)

- University of Virginia Phun Physics Show: *http://phun.physics.virginia.edu/demos* (See Bell jar/shaving cream in vacuum; Collapsing drum; Magdeburg hemispheres; Marshmallow man.)

- Wake Forest University: Physics of matter: Pressure demonstration videos: *www.wfu.edu/physics/demolabs/demos/avimov/bychptr/chptr4_matter.htm*
- Whelmers #21 Balloon (in Bottle) Vacuum: *www.mcrel.org/whelmers/whelm21.asp*
- Wikipedia: Cognitive load theory: *http://en.wikipedia.org/wiki/Cognitive_load_theory* and Zone of proximal development: *http://en.wikipedia.org/wiki/Zone_of_Proximal_Development*

Answers to Questions in Procedure, steps #1–#4

1. The wood flips in a somersaulting motion just as most people would predict.

2. A likely response will be that the stick will again fly up but that this time the stick will either take the paper with it or rip the paper.

3. When the newspaper is placed over the wood, the wood breaks right at the edge of the table if the demonstrator strikes hard rather than slowly pressing down on the wood. In the latter case, the newspaper is lifted up. By smoothing the paper firmly against the table and removing air from underneath, you create a situation where the wood sticking up is pushing against the weight of a column of air that extends to the outer limits of the Earth's atmosphere. Inertia causes the paper to remain at rest and the rapidly moving end of the stick to keep moving, which it does by snapping at the point where it extends just beyond the table. Conversely, if the wood stick is pushed down slowly, the air that seeps in underneath the paper can exert pressure upward to counterbalance the air pressure on top of the paper and the stick can easily lift the paper up without snapping. The relevant explanatory facts are as follows: air pressure = force/area = weight of the column of air/surface area. At sea level, air pressure = 10 newtons/cm^2 or 14.7 lbs/in^2. (*Note:* 10 N = weight of a 1 kg mass at sea level.)

4. The calculation of the effective weight of the column of air that is pressing down on the surface area of a single piece of newspaper is as follows: surface area = 61 cm × 53 cm = 3,233 cm^2 and weight = 3,233 cm^2 × 1 kg/cm^2 = 3,233 kg (technically, kg is a unit of mass, not weight) or nearly 7,113 lbs!

3D Magnetic Fields: Making Meaningful Connections

COW MAGNET

MINERAL OIL

Expected Outcome

A sealed, transparent container filled with a clear, colorless oil and several tablespoons of iron filings is shaken and a

IRON FILLINGS

cylindrical magnet is suspended in the middle of the container. Beautiful magnetic field lines are made visual as the iron filings are suspended in a specific 3D pattern that changes when influenced by outside magnetic fields.

Science Concepts

Magnetic fields are 3D, yet many textbook representations and projected demonstrations that use iron filings visually suggest they are 2D. Michael Faraday, the brilliant, self-educated (but somewhat mathematically illiterate) English physicist and chemist, was the first to propose (in 1844) that invisible, 3D "lines of force" could account for the mysterious action-at-a-distance effect of magnets. He used iron filings and sketch drawings to visualize these otherwise invisible magnetic force fields. (*Note:* As developed in the Procedure section, this activity is a quicker, more visually appealing demonstration [used as part of an Explain or Elaborate phase] than an in-depth science inquiry activity. Accordingly, no science inquiry questions are included.)

Science Education Concepts

Modern cognitive psychologists have confirmed what ancient philosophers suggested, namely, that learning is about making cognitive connections between prior understanding and new experiences. The human brain is designed to "make sense" by wiring and rewiring neural networks that encode memory and understanding. Teachers' use of integrated curriculum, instruction, and assessment plans that are organized around big ideas and conceptual themes help students truly understand (rather than merely memorize) science. At the same time, students develop their metacognitive capabilities (i.e., awareness of and ability to monitor and intentionally direct their thinking processes). Meaningfulness, the opposite of mindlessness, is both a prerequisite for and product of real learning. Disciplinary content is meaningful if students perceive that it "makes sense" in light of their prior knowledge and is relevant to their interests.

Various types of graphic organizers have been developed to help learners make their otherwise invisible thinking visible to both themselves and their teachers (Mintzes, Wandersee, and Novak 1998; also, Internet Connections: IHMC and Visible Thinking). The graphic organizer called concept mapping, in particular, is an excellent teaching and learning technique for researchers, science teachers, and students (Good, Novak, and Wandersee 1990). Teachers can use concept maps to help them design instructional units, while student-developed concept maps help teachers to visualize, assess, and monitor students' preinstructional and evolving sense of conceptual relationships.

As a simplistic *visual participatory analogy*, one can compare three things: static images of ever-changing neural networks ("neurons that fire together, wire together"), snapshot-in-time concept maps, and 3D magnetic fields made visible. Connections and organized structures or patterns are the common key in all three of these cases. Similarly, big ideas in science (e.g., atoms, cells, evolution, plate tectonics, and the conservation of energy) enable scientists to see and explain the underlying ordered patterns and unity ("cosmos") in the seemingly endless diversity ("chaos") of natural phenomena.

Materials

- An inexpensive homemade version of a 3D magnetic field box can be constructed as follows. Nearly fill a clear plastic, 1 L soda bottle with colorless mineral or silicon oil and several teaspoons of iron filings (or small pieces of cut-up steel wool). Insert a cylindrical magnet wrapped into a sealed test tube (or wrap it in a small, sealed plastic baggie) and suspend it by a thread to the screw cap of the soda bottle. Pairs of magnets on the outside of the bottle can be used to study the effect of two opposite poles versus the effect two identical poles. Both the Exploratorium Snackbook and Indestructables websites discuss this model—see Internet Connections.

- Strong, cylindrical "cow magnets" are available from most feed stores (ranchers "feed" these to cows to prevent "hardware disease"—that is, intestinal lacerations caused when cows inadvertently eat iron scraps. They typically are Alnico magnets, an iron alloy containing Al-Ni-Co). The use of oil prevents the oxidation of the iron filings (which would cause the bottle to collapse over time as oxygen combines with iron to form iron oxide rust) and helps suspend the iron filings for a better visual effect.

- Alternatively, complete magnetic field kits are commercially available from several sources:
  - Arbor Scientific. *http://arborsci.com.* 800-367-6695. Magnetic Field Observation Box (with cylindrical cow magnet and iron filings suspended in silicone oil in a 4 in. × 2 in. × 2 in. acrylic box). P8-8001. $105. Expensive, but visually stunning when placed on an overhead or light board.

Safety Note

Students and teacher should wear safety glasses or goggles during this activity.

- Edmund Scientific. *http://scientificsonline.com.* 800-728-6999. The Mysterious Magnet Tube #3052976. $14.95. A sealed 4 in. × 3 in. cylindrical acrylic air-filled viewing tube with 1.3 oz. of iron filings and an Alnico cow magnet.
- The Magnet Source. *www.magnetsource.com/Solutions_Pages/ cowmags.html.* 888-293-9399. Sells a wide variety of magnets, including cow magnets.
- The PhysLink.com Science eStore. *www.physlink.com/eStore/ cart/Magnets.cfm.* 888-438-9867. 3D Magnetic Field Tube (air-filled cylinder with an Alnico magnet). #10147. $12.95.
- When working with teachers, sample images of neural networks (see Internet Connections: Neural Networks) may be useful to help them visualize the analogy between magnetic fields and neuron-neuron networking and communication.

Points to Ponder

Science is constructed of facts, as a house is of stones. But a collection of facts is no more science than a heap of stones is a house.

—Jules Henri Poincaire, French mathematician and astronomer (1854–1912) *in La Science et l'Hypothese* (1908)

One of the best selling secondary chemistry texts in the United States contains over 700 pages and weighs about 1.5 Kg…. This is far more information than any student can be expected to learn and understand in one year…. Today's chemistry textbooks, and probably the course based on them as well, serve to turn interested students away from chemistry.

—Linus Pauling, American chemist (1901–1994) in *The Science Teacher,* September 1983.

Enormous amounts of information are available, including however, very little reliable data on what it all means.

—Ashleigh Brilliant, American humorist (1933–) *in All I Want Is a Warm Bed and a Kind Word and Unlimited Power: Even More Brilliant Thoughts* (1985)

Procedure

When Working With Teachers (and With Students, if Desired)

1. Share one or more of the quotes by Poincaire, Pauling, and Brilliant while holding up a science textbook. Discuss how a person's true understanding of the material in this book would depend less on memorizing all the separate, individual facts (or "individual trees") and more on understanding how the facts fit together to form conceptual patterns of meaningful relationships (the "forest"). The Gestalt principle of synergy—the whole is

greater than the sum of the parts—should be a guiding principle in teaching and learning science. The idea of emergent properties also applies to all levels of matter organization from subatomic particles → atom → molecules → cells → ... universe.

2. Shake up the container so that the iron filings become temporarily suspended in the oil. Draw an analogy between the many separate, chaotically arranged filings and misguided efforts of students to mindlessly memorize isolated, seemingly unrelated science factoids. Insert the cylindrical magnet into the container as an organizing factor (or visual analogy for the AAAS Benchmarks' "common themes," the National Science Education Standards' "unifying concept or process," or any big idea theory in science). This phenomenon is visually dramatic when the container is illuminated from below by placing it on a light board or overhead projector. Other magnets placed inside or outside the container will disturb the original pattern and create a new, modified 3D pattern analogous to the way that new experiences modify prior conceptions.

   Challenge the learners to consider ways that this analogy might not be a good one for learning science (e.g., the analogy might suggest that the organizing theme or theory could be simply given to and passively absorbed by the student). On the other hand, the student's mind could be considered as playing the active, organizing role of the magnet. In any case, compare and contrast the beautiful 3D magnet patterns to sample images of neural networks (see Internet Connections and search Google).

When Working With Students

1. After students have experienced a variety of Engage- and Explore-phase activities related to magnetism, move on to the Explain (or Elaborate) phase and talk about students' discoveries about invisible magnetic fields. You could discuss and demonstrate Oersted's 1819 discovery of the 3D magnetic fields around wires with flowing current. This would be a lead-in to Faraday's groundbreaking work, which would help students to visualize, conceptualize, and use (electro) magnetic forces.

184

2. If multiple, cheaper, homemade models are assembled, students can play with changing the structure of the 3D field by placing other magnets at various locations outside the container. Also, if two cylindrical magnets are inserted from opposite ends, students can compare and contrast the patterns formed by N ←→ N, S ←→ S, and N →← S arrangements. Computer simulations (see Internet Connections: PhET) can be used to help students visualize the invisible magnetic (and electric) fields.

Debriefing

Both teachers and grades 5–12 students need to understand that learning science is not about students mentally archiving unrelated facts for later retrieval on a test. Rather, it is about making sense of—or making meaning from—natural phenomena in light of a limited number of conceptual themes and far-reaching scientific theories. This demonstration can introduce students to how to use various kinds of graphic organizers, such as concept mapping. Concept maps attempt to represent at a macroscopic, simplified level the internal, linked hierarchal relationships among mental concepts. As such, concept maps might be considered loosely analogous to the neurological networks that encode conceptual understanding and meaning (see Internet Connections: IHMC). The old folk song "The Green Grass Grows All Around" (lyrics at *www.kididdles.com/lyrics/g012.html*) works well as an auditory analogy for emphasizing interrelationships and connectedness of concepts (also see Extension #3). Teachers may wish to explore the research on student misconceptions about magnetism (e.g., see Driver et al. 1994, Chapter 16, as a starting point).

Extensions

1. *Magnetic Fields Matter in Movies and Medicine.* The science-fiction movie *The Core* (2003) has a dramatic opening sequence that shows various discrepant events that could be caused by a disruption in the Earth's invisible 3D electromagnetic field. (Like most science fiction movies, this one contains both "good" and "bad" science.) Magnetic resonance imaging (MRI) is a noninvasive diagnostic technology that has revolutionized medicine by making visible what is otherwise invisible.

2   The book *How Everything Works: Making Physics Out of the Ordinary* by Louis A. Bloomfield and the HowStuffWorks website (see Internet Connections) can be used to explore mag-lev trains, the aurora borealis, and other real-world applications of magnetic fields.

3. *Magnetic Construction Toy Sets.* Toy stores sell kits consisting of magnetic bars and steel balls that can be arranged to form a large number of different shapes from the exact same set. The instructor can distribute the same kit to several different teams and ask them to construct a 3D figure of their own choosing. The different figures created can be used as a visual analogy for how students, given the same curriculum and instruction, can form different conceptual understandings (especially if the curriculum is ill-defined). These unique constructions can also be connected to the idea of concept maps (see Internet Connections).

4. *Musical Memory as Mnemonics.* Music and instruments are traceable back to the earliest stages of human civilization (long before the development of permanent cities). Advertisers make use of musical memory in their product jingles to help consumers keep their products (versus those of their competitors) in mind. Similarly, both published and student-written songs can help students learn key principles of science. The website Neuroscience for Kids (see Internet Connections) contains a number of "brain songs" that emphasize the idea of neural connections and meaning-making. The songs are based on popular folk songs: "The Dendrite Song" ("Oh My Darling Clementine"), "I've Been Working on My Neurons" ("...the Railroad"), and "Home, Home in the Brain" ("...on the Range"). Such songs can serve as models to help students learn how to create their own science lyrics using melodies from popular songs as a mnemonic or organizing tool that aids in comprehension, retention, and recall. Physical movement can be added to the rhythm and rhyme to develop procedural (or how-to) and episodic memory to supplement and strengthen declarative memory networks (in this case, perhaps, participating in a "funny dance" in a classroom setting). See also Activity #12 in this book.

Internet Connections

- Arbor Scientific's Cool Stuff Newsletter: *www.arborsci.com/Cool Stuff/Archives3.aspx*. (See magnetism listing and newsletter #17.)

- Exploratorium Snackbook: Magnetic lines of force (homemade 3D magnetic field bottle): *www.exploratorium.edu/snacks/magnetic_lines/index.html*

- HowStuffWorks: (1) Aurora borealis: *http://science.howstuffworks.com/question471.htm* (2) Brain: *http://health.howstuffworks.com/brain.htm* (3) Mag-lev trains: *http://science.howstuffworks.com/maglev-train.htm*

- Indestructables: Make your own homemade 3D magnetic field viewer: *www.instructables.com/id/3D-Magnetic-Field-Viewer*

- Institute for Human and Machine Cognition (IHMC): Concept mapping tools (free software): *http://cmap.ihmc.us* (Dr. Joseph Novak, the originator of concept mapping, is affiliated with IHMC.)

- International Mind Brain and Education Society: *www.imbes.org*

- JAVA Applets for Physics: Magnetic field of a bar magnet: *www.walter-fendt.de/ph14e/index.html*

- Magz Magnetic Construction Toy: *www.magz.com*

- Neural Networks: Sample images: *http://chemistry.caltech.edu/~fucose/Neural%20Coonections.htm* and *http://brainenrichment.blogspot.com/2008/02/rss-reminds-me-of-how-brain-works.html*

- Neuroscience for Kids: *http://faculty.washington.edu/chudler/neurok.html*. Select experiment: Brain songs.

- PhET Interactive Simulations: Magnets and electromagnets: *http://phet.colorado.edu/simulations/sims.php?sim=Magnets_and_Electromagnets*

- Royal Institution of Great Britain: Faraday: *www.rigb.org/rimain/heritage/faradaypage.jsp*

- Society for Neuroscience, Educational Resources: *www.sfn.org/index.cfm?pagename=PublicEducationOutreach_NeurosciEdu Resources*

- University of Virginia Physics Department: Investigating magnetic fields (in 2D): *http://galileo.phys.virginia.edu/outreach/8thGradeSOL/MagneticFieldsFrm.htm*

- Visible Thinking (a Harvard University, Project Zero research program): *http://pzweb.harvard.edu/vt/VisibleThinking_html_files/VisibleThinking1.html*

- Wake Forest University: Electromagnetism videos including a 3D compass: *www.wfu.edu/physics/demolabs/demos/avimov/bychptr/chptr8_eandm.htm*

- Wikipedia: (1) Chemical synapse: *http://en.wikipedia.org/wiki/Synapses* (2) Concept mapping: *http://en.wikipedia.org/wiki/Concept_map* (3) Cow magnet: *http://en.wikipedia.org/wiki/Cow_magnet* (4) Earth's magnetic field: *http://en.wikipedia.org/wiki/Magnetic_field_of_earth* (5) Magnetic resonance imaging (MRI): *http://en.wikipedia.org/wiki/MRI* (6) Magnets: *http://en.wikipedia.org/wiki/Magnets* (7) Michael Faraday: *http://en.wikipedia.org/wiki/Michael_Faraday* (8) Neural networks: *http://en.wikipedia.org/wiki/Neural_networks*

Activity 18

Electric Generators: Connecting With Students

Expected Outcome

A hand-powered electric generator converts mechanical energy into direct current (DC) electricity that can be used to power a small lightbulb, run an electric motor (i.e., a second identical generator), and do other tasks that require electricity. Electrical polarity can be reversed by switching the lead wires or by cranking the generator handle in the opposite direction.

Science Concepts

Relative motion between a magnet and an electrical conductor (e.g., a coil of wire) will produce an electric current (and an accompanying magnetic field) that can then be converted into other forms of energy. Electrical generators and motors were independently invented in the 1830s by the American physicist Joseph Henry and English physicist Michael Faraday. Neither scientist filed a patent for these inventions, as they believed the benefits of readily available electric power should benefit all humanity. Handheld generators are a safer and cheaper alternative to plug-in AC power sources for doing many classic experiments with current electricity.

Science Education Concepts

Teaching involves social interactions and cognitive connections between a teacher and one or more learners who interact with each other and with "live" science phenomena. "Translating" published science standards into a teacher's curriculum-instruction-assessment plan (using the teacher's words and instructional activities), then to students' senses, and finally to students' diverse conceptual frameworks is a complicated series of events and can lead to missed connections and miscommunication. As such, plans need to be based on an iterative design cycle of scaffolded lessons (e.g., BSCS's 5E Teaching Cycle; see Appendix B) that are adjusted in light of ongoing diagnostic and formative assessment. This feedback data is used to determine what is actually being "caught by the learner" (to use a somewhat passive and misplaced metaphor) rather than focusing only on what has been taught by the teacher. Meaningful learning experiences allow students to reinforce and "rewire" the cognitive connections that make up their neuro-electrochemically based conceptual networks as well as to fill-in conceptual holes (see, for example, Brooks and Brooks 1999; Jensen 1998; Mintzes, Wandersee and Novak 1998). Neuroplasticity and synaptic connections are the basis for memory, learning, and creativity (or having "bright" ideas). The electrical analogy is developed and critiqued in Teacher Procedure (p. 192).

Materials

- Minimum of two transparent, hand-cranked electric generators with external leads (to demonstrate the generator → motor connection). Suppliers include the following:
  - Arbor Scientific. *http://arborsci.com.* 800-367-6695. Genecon. P6-2631 $54 (produces up to 5 V DC and 200 mA usable current). See also accessories and lab manuals. Arbor Scientific also sells the similar but cheaper Hand Generator. P6-2550. $19.95 (up to 7.5 V output).
  - Educational Innovations. *www.teachersource.com.* 888-912-7474. Hand-Cranked Generator (and flashlight bulb). #GEN-100. $21.95 (up to 12 V DC).

Both of these suppliers also sell single purpose Dynamo Hand-Powered Flashlights and electromagnetic (or Faraday) flashlights. These products are also available in drugstores and department stores (and use either capacitors or rechargeable batteries to store a charge). The teacher can use generators to light standard flashlight bulbs, lights cut from low cost Christmas strands (leave several inches of wire on each side of the bulbs), or light-emitting diodes (LEDs).

Points to Ponder

I know you believe you understand what you think I said. But, I'm not sure you realize that what you heard is not what I meant.

—Anon.

There is no plea that will justify the use of high tension and alternating currents, either in a scientific or a commercial sense … not only on account of danger, but because of their general unreliability and unsuitability for any general system of distribution.

—Thomas Alva Edison, American inventor (1847–1931)

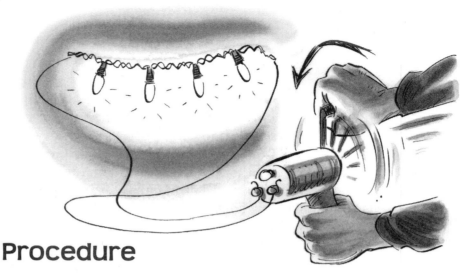

Procedure

When Working With Teachers

If possible, use several setups so multiple volunteers can get a direct hands-on feel for the phenomenon. Otherwise, two generators are sufficient for demonstration purposes. (See answers to questions in steps #1 and #2 on pp. 197–198.)

1. Create a visual analogy for interconnected teaching and learning by using the hand-cranked generator [teacher] attached to a flashlight bulb [student]. Demonstrate that when the teacher puts energy into the system, the lightbulb lights up. Ask: What would be the effect of turning the handle with more speed or adding additional lightbulbs (in series)? What would happen if one of the two contact points with the bulb were removed or broken? If time permits, ask: How are these demonstrations similar or analogous to the relationship between teacher and student? What are the flaws with this analogy as a model for teaching and learning?

2. Ask: (a) How can we use the materials at hand to make a better visual participatory analogy for the interactive, minds-on nature of teaching and learning? (b) What could we add to this system so that it modeled residual, lasting learning on the part of the student?

When Working With Students

The teacher may want to do the Engage-phase demonstration variations in Teacher Procedure, step #1, above (see also Activity #23 in this book) to raise questions and motivation prior to students' hands-on explorations. Lab manuals that accompany commercially purchased generators contain a number of experiments to use before, during, or after Explain-phase interactive discussions. These include experiments that duplicate those done with batteries and/or more dangerous AC sources of power (e.g., the effects of current on a compass, electromagnets, conductors versus insulators, electrochemistry, series and parallel circuits, mystery circuit boxes, variable resistors, short circuits and fuses, the thermoelectric effect, and Ohm's law). Sample introductory Engage-phase questions include the following:

(See answers to questions in steps #1–#4 on pp. 198–199.)

1. In a darkened room, ask students, Where does the energy to light the room come from? Flip the switch and ask, What energy conversions are involved in powering the lightbulbs? Is energy created or destroyed in the process? Simulate part of this process with a demonstration of the mechanical energy or motion input → generator → electricity output → light connections.

2. What is the relationship between an electric generator and a motor? How can we demonstrate this relationship with the two generators?

3. What is the difference between direct current (DC) and alternating current (AC)? Which type of current is produced by public utility electric power plants and why? On what basis was this decision between competing forms of electrical current made? (These questions can be explored via a science-technology-society research project and reported on later in the unit.)

4. What other questions about electrical energy, power, and circuits could we explore with a generator (or motor)?

Debriefing

When Working With Teachers

Teaching is about helping students strengthen old and/or make new cognitive and emotional connections (and often involves "rewiring" prior neural electrochemical connections for an expanded sense of meaningfulness and usefulness). Unlike this simple electrical circuit model, each of the brain's 100+ billion neurons can connect to a matrix of thousands of other neurons. It is this ever-evolving matrix that forms the biological, electrochemical, and structural foundations of learning and

memory. In a sense, teachers are "brain surgeons" (or "electricians") who work from the outside on "patients" (or "houses") who (or which) are "in charge" and actively involved in performing their own "self-surgery" or "rewiring." Learning at external, macroscopic or observable levels and at internal, microscopic or hidden levels is all about making connections. This process is aided by well-designed, adaptive curriculum-instruction-assessment plans (see Internet Connections: BSCS 5E cycles and Constructivism) that (1) help students see the information to be learned as meaningful and relevant to current and prior experiences and (2) encourage two-way teacher ←→ student interactions that provide ongoing feedback to both collaborators. Use the anonymous quote to highlight the multiple places where communication breaks down and misconceptions can occur. Teachers can explore the large body of published research on student misconceptions about electricity (e.g., see Driver et al. 1994, Chapter 15, as a starting point).

When Working With Students

Most students have a very limited sense of where electricity comes from (beyond batteries and wall sockets). They simply flip a switch, turn a dial, or press a "remote," and electricity is there, available to do their bidding in a wide variety of devices and consumer products. In fact, the typical American teen uses more total energy (i.e., home heating and AC, internal combustion engines, and plug-in and battery-powered electrical devices) than the combined human slave and animal "horsepower" that was available to the richest of ancient citizens. Understanding the energy crisis and the related environmental crisis requires that students have a direct, hands-on sense of where electricity comes from and what is required to produce it. Work with commercial and/or student-constructed generators and motors would typically be found in a unit on electromagnetism. Some utilities lend out bike-powered electric generators that can power large-wattage household products (e.g., radios, TVs, floodlights, and heaters). Computer simulations (see Internet Connections: PhET) can also be used to help students visualize how electrical generators convert kinetic energy into electricity to light a lightbulb.

It is worth pointing out, as we see from Edison's quote, that Edison's vision of an electric power system was based on DC power plants located very close to their customers. He actively fought against the idea of AC power plants and high-tension transmission lines. He even used his political clout to ensure that the first prisoner condemned to death by electrocution was "Westinghoused" with alternating current! (See Internet Connections: Death, Money, and the History of the Electric Chair and Snopes. Com.) It is the AC system (as championed by his competitor, George Westinghouse) that accounts for nearly all the electric power generated in this country. Even local, on-site, solar-generated electrical energy and battery-powered cars connect into the AC power grid. This example demonstrates that great scientists and inventors are not immune from personal biases (often based on personal financial gains), scientific misconceptions, and limited visions for the future!

Extensions

1. Both Arbor Scientific (P8-8300) and Educational Innovations (#SS-11) sell the World's Simplest Motor kits for under $5. The kits allow students to build a motor that will run up to five hours on a single D-cell battery. The Internet Connections contain a variety of plans for constructing low-cost, homemade generators and motors.

2. The Official M. C. Escher Website (*www.mcescher.com*) is a comprehensive source for information about this artist and his unusual visions. His lithograph *Bond of Union*—showing the interconnections between the minds of a male and female (and available on a poster and a T-shirt)—serves as a great visual metaphor for interactive, collaborative, mutually rewarding teaching and learning.

Internet Connections

- Amateur Science's Ultra-Simple Electric Generator: *http:// amasci.com/amateur/coilgen.html*

- Arbor Scientific's Cool Stuff Newsletter: *www.arborsci.com/Cool Stuff/Archives3.aspx.* (See Electricity: Homopolar motor, motors, and generators and Newsletter #38 demonstrations.)

- BSCS 5E Instructional Model: Origins, effectiveness and applications: *www.bscs.org/pdf/5EFull Report.pdf* (65 pages) and *http:// bscs.org/pdf/bscs5eexecsummary.pdf* (19 pages).

- Constructivism links: *http://carbon.cudenver.edu/~mryder/ itc_data/constructivism.html*

- Death, Money, and the History of the Electric Chair: *http://inventors.about.com/od/hstartinventions/a/Electric_Chair.htm*

- Exploratorium Snackbook: Motor effect and stripped-down motor (designs): *www.exploratorium.edu/snacks/motor_effect* and *www.exploratorium.edu/snacks/stripped_down_motor*

- How Electricity Works: *http://science.howstuffworks.com/electricity.htm*

- Java Applets for Physics: Generator: *www.walter-fendt.de/ph14e/index.html*

- Nexus Research Group: Motors and generators: *www.nexusresearchgroup.com/fun_science/motors.htm*

- PhET Interactive Simulations: Faraday's law and electromagnetic lab: (1) *http://phet.colorado.edu/simulations/sims.php?sim=Faradays_Law* (2) *http://phet.colorado.edu/simulations/sims.php?sim=Faradays_Electromagnetic_Lab* (3) Generator (water powered): *http://phet.colorado.edu/simulations/sims.php?sim=Generator*

- Science Project: Make an electric generator: *www.scienceproject.com/A/projects/KITWG/index.asp*

- Snopes.Com: Urban legends: Edison and the electric chair (it's true!): *www.snopes.com/science/edison.htm*

- University of Iowa Physics and Astronomy Lecture Demonstrations (video clips): Electricity and magnetism: *http://faraday.physics.uiowa.edu* (seven demos on motors and generators)

- University of Virginia Physics Department: HOEs with electromagnets: *http://galileo.phys.virginia.edu/outreach/8thGradeSOL/ElectromagnetFrm.htm*

- WikiPedia: (1) Constructivism: *http://en.wikipedia.org/wiki/Constructivism* (learning theory and teaching methods) (2) Electrical generator: *http://en.wikipedia.org/wiki/Electrical_generator* (3) Michael Faraday: *http://en.wikipedia.org/wiki/Michael_Faraday* (4) Joseph Henry: *http://en.wikipedia.org/wiki/Joseph_Henry*

Answers to Questions in Procedure

When Working With Teachers, steps #1 and #2

1. The generator-teacher → lightbulb-student analogical demonstration shows how a teacher output becomes a student input that, in turn, generates a response by the student (unless the lightbulb-student is "dead"). Unless you demonstrate that there needs to be a connection *back* to the generator-teacher (i.e., by disconnecting one of the two leads), many teachers might miss this scientific and pedagogical reality. The fact that the lightbulb lights up only when you have a complete, closed circuit suggests that there needs to be some form of communication from the teacher to the student *and back* (although the feedback itself is not visible—only the lightbulb-student's reaction).

    However, this visual participatory analogy is flawed in several ways:

*   It all depends on the generator-teacher; the lightbulb-student has a somewhat passive, receiving-and-reacting role.

*   Unless the lightbulb is "dead," the effect of the teacher's effort is immediately apparent; student learning, on the other hand, is rarely so immediate or so easy to detect and assess.

*   There is no residual effect on the student, who stays "lit up" only as long as the teacher is working.

*   The flow of energy appears to be in one direction only—from generator-teacher to lightbulb-student—and the process cannot be initiated by the lightbulb-student.

*   Greater speed on the part of the teacher results in the bulb glowing more brightly, but speeding up in the classroom by talking faster or racing through a lesson is rarely a good idea.

*   The teacher feels more resistance (i.e., the crank is harder to turn) when additional "students" are added (i.e., additional lightbulbs are added in series or even a single "brighter" bulb is used). This last point runs counter to the many advantages of learning in a social environment, where peers can enrich the opportunities to learn (and the teacher feels a certain ease when working with a particularly "bright" student).

2a. If two generators are attached together and one is cranked, it will generate electricity that will be converted back into motion in the rotation of the crank in the second generator (now turned into a motor). With this model, information (i.e., electrical energy) can flow back and forth in either direction between the "teacher" and the "student," just as should occur in a classroom where asking and answering questions, generating and critiquing ideas, and other interactive processes provide formative feedback to both teacher and students. The direction the handle of the motor turns changes in response to the direction the generator's crank is turned and the way the leads are connected.

2b. A capacitor or a rechargeable battery could be used to represent a student who retains energy for later use. For example, Faraday flashlights and hand-cranked flashlights store the energy for later use. Of course, both will eventually "run out"—something that is not true with actual learning, which results in the rewiring of internal neural networks that encode meaningful learning. (In fact, neuroscientists indicate that when it comes to brain and memory, it's "use it or lose it!")

When Working With Students, steps #1–#4

1. The energy transfer sequence is as follows: chemical potential energy of a fossil fuel → heat energy → mechanical energy (motion of a steam-turned turbine) → electromagnetic energy (in a generator) → light (+ heat). Energy is always conserved, but it is changed in form, and ultimately, previously useable energy is dissipated as "waste" heat of random molecular motion.

2. An electric generator and a motor are essentially the same device, except that in the case of a generator, mechanical energy is an input and electricity is an output; the reverse is true with a motor. If the output wires of one "generator" are attached to the input wires of the "motor," hand cranking the generator will cause the crank on the motor to rotate.

3. Direct current can be visualized as "flowing" in one direction in a continuous loop from the generator to the motor and back. Alternating current reverses direction 60 times per second (60 Hz) in the United States. Power plants use AC because it is easier to

step up its voltage for more efficient transmission over long distance, high-tension wires and then use step-down transformers to drop it back to 120 V for household uses. Reducing the loss of usable energy in transmission is an area of ongoing research. Experiments verified by other scientists are the gold standard of provisional truth in science.

4. Other possible experiments are mentioned in the introduction to the Student Procedure.

Static Electricity: Charging Up Two-by-Four Teaching

Expected Outcome

A 6–8 ft., 2 in. × 4 in. board balanced on a large watch glass rotates toward a balloon that has been electrically charged by being rubbed against the instructor's hair.

Science Concepts

Static electricity (or triboelectricity) is created when two different insulating materials are rubbed together, creating friction that allows electrons to shift from one material (that becomes positively charged) to the other (that becomes negatively charged). Negatively or positively charged objects can attract neutral objects by inducing internal, localized displacement of mobile, negative charges in the neutral object. For two charged objects, the law of charges states that opposite charges attract and like charges repel.

Science Education Concepts

When teachers use a funomena instructional approach at the beginning of a lesson, they are activating students' attention and their cognitive processing. Multisensory Engage and Explore activities (see Appendix B for an explanation of the 5E Teaching Cycle) provide multiple contexts to ground subsequent Explain activities, when terms and concepts are formally introduced. Teachers should let students ask questions first before prematurely helping them construct answers. Heavy-handed, 2 × 4 teaching leads to static and bored students. Effective science teaching extends beyond the two (covers of the textbook) by four (walls of the classroom); it brings real-world applications into the classroom and takes students out of the classroom (e.g., field investigations, virtual tours, and simulations) to bring about meaningful learning.

Materials

- 1, 6–8 ft., 2 in. × 4 in. board (or broomstick)
- 1 large, ~15 cm diameter watch glass (or a curved ice cream scoop)
- 1 round balloon
- *Alternative:* The board could be suspended from the ceiling by a heavy-duty filament (or metal wire) attached to the board's center of mass.

Note: Static electricity experiments are best done when the relative humidity is low, as high humidity limits the buildup of static charges that are necessary to cause noticeable effects. Also, oily hair and the use of hair conditioners will adversely affect the buildup and transfer of electric charge from human hair to a balloon.

Points to Ponder

One had to cram all this stuff into one's mind for the examinations, whether one liked it or not. This coercion had such a deterring effect ... that after I had passed the final examination, I found the consideration of any scientific problem distasteful to me for an entire year.... It is, in fact, nothing short of a miracle that the modern methods of instruction have not yet strangled the holy curiosity of inquiry.... It is a very grave mistake to think that the enjoyment of seeing and searching can be promoted by means of coercion and a sense of duty.

—Albert Einstein, German-born American physicist (1879–1955)

If scientific education is to be dealt with as mere bookwork, it will be better not to attempt it....The great peculiarity of scientific training, that in virtue of which it cannot be replaced by another discipline whatsoever, is ... drawing conclusions from particular facts made known by immediate observation of Nature.... You must not be solicitous to fill him with information, but you must be careful that what he learns he knows of his own knowledge. Don't be satisfied with telling him.... Tell him that it is his duty to doubt until he is compelled, by the absolute authority of Nature, to believe that which is written in books.

—Thomas Henry Huxley, British evolutionary biologist (1825–1895) in *Science and Education* (1899)

Procedure

Optional: When Working With Teachers

Begin the session with a 1–2 min. parody of a "bad lecture" that uses foreign terminology, no concrete demonstrations of the underlying concepts, and lots of factoids to be memorized—for example, "Today we are going to talk about the electrostatic series that runs from ma-

terials that tend to assume a positive polarity to those that assume a negative polarity such as the series from rabbit's fur, glass, mica, nylon, wool, cat's fur, silk, paper, cotton, wool, Lucite, sealing wax, amber, polystyrene, polyethylene, rubber balloon, sulfur, celluloid, hard rubber, vinylite, Saran wrap, and Teflon. The farther apart the materials are on the list, the higher the charge separation will be when they are rubbed together. Triboelectricity was first studied by…."

When Working With Teachers and Students

(For answers to questions in steps #1–#4, see p. 209)

1. Hold up the long 2 in. × 4 in. board. Ask: How can I use this board to activate attention and catalyze cognitive processing in a way that students will not forget? What can this board teach us about the mysterious, invisible particles and forces that make up the universe?

2. Place a large watch glass (a watch glass is a circular, concave piece of glass used in chemistry as a surface to evaporate a liquid, to hold solids while being weighed, or to serve as a cover for a beaker) on a demonstration table or elevated box (a curved ice cream scoop can be used in place of the watch glass). Balance the long board with the 4 in. side facedown on the watch glass. Lightly move the 2 in. × 4 in. board with your hand to show that it can freely rotate about its midpoint. Ask: How might I move this board without touching it either directly or indirectly (via another object that does touch it)? After eliciting comments, proceed to blow with great enthusiasm on the board to make it pivot. Ask: Do you think a science teacher's brain power exceeds that of the typical human brain, which uses energy at a rate somewhat less than that of a 25 W lightbulb? Can I use my high energy "brain electricity" to exert a mind-over-matter force to move this board?

3. Rub a round balloon against your hair as if you are somewhat absent-mindedly trying to think of a way to move the board with "mind power." Hold the negatively charged balloon adjacent to, but not touching, the balanced 2 in. × 4 in. board; the neutral board will be attracted toward the balloon. (*Note:* Human hair, like cat and rabbit fur, will lose electrons to rub-

ber or latex and become positively charged as the rubber or latex becomes negatively charged.) Continue to pull the balloon away and the board will pivot on its axis in a complete circle. Ask: How might I use this balloon to stop and even reverse the direction of the board's motion?

4. Ask: Is the mysterious, action-at-a-distance force something unique to science teachers, or is it a universal force that could be employed by anyone? How does this force relate to invisible particles of matter? Is it always attractive, or if not, under what circumstances might it repel other objects? If time permits, use a second uncharged balloon to allow a volunteer to replicate the discrepant event. If desired, balance a broomstick, an aluminum level, or meter stick on the watch glass. Also, if you rub the balloon against Saran Wrap, Teflon, or another material lower than rubber on the electrostatic series, the balloon will lose electrons and becoming positively charged. A positively charged balloon will also attract the neutral 2 in. × 4 in. board via induction. Either a negatively or a positively charged balloon will attract neutral objects.

Debriefing

When Working With Teachers

Discuss the importance of giving students a variety of multisensory experiences when investigating scientific phenomena. Science instruction that limits students' experiences to words that exist between the two covers of the textbook and experiences bounded by the four walls of a standard classroom (i.e., 2 × 4 teaching) is almost guaranteed to produce bored students. Share and discuss the Einstein quote with respect to the deleterious effect of coercive methods of teaching; consider an analogy to the "pedagogical approaches" of a Jules Verne– versus a Hitler–type teacher. Use the Huxley quote to compare the relative power of the "carrot" of natural phenomena and minds-on and hands-on explorations versus the "stick" of textbooks, teacher talk, and tests when the goal is to motivate and sustain students' effortful learning. Teachers can explore cognitive science research that counters popular "brain-based" learning "myths" in the Ask the Cognitive Scientist column that appears in *American Educator* (Internet Connections).

When Working With Students

This activity works as an "edu-taining" discrepant introduction (or Engage phase) to a unit on static electricity. *Do not* attempt to prematurely Explain the phenomenon, but rather use the discrepant-event demonstration to create a need-to-know. Challenge the students to raise specific questions about static electricity that they would like to Explore. Brainstorm examples of static electricity in their lives (e.g., photocopy machines and laser printers, static cling in drying clothes, static shocks during the winter, lightning) that the class will ultimately be able to explain. Explore-phase activities would involve hands-on discovery of, for example, the law of charges, the idea of charging by induction, and the electrostatic series. Computer simulations can then be used to help visualize the movement of invisible charges (see Internet Connections: Concord Consortium and PhET).

Extensions

1. *Electrifying Exploration.* Charged balloons can be "commanded" to "sit on" imaginary, invisible book shelves attached to the walls (or ceiling); bend a stream of water toward their charged side; pick up small pieces of paper; attract empty aluminum soda cans placed on their sides and allowed to roll (in "triboelectric track races"); momentarily light up a fluorescent tube held in your hand when brought in contact with the metal prongs, and so forth. If antistatic, fabric softener sheets used in clothes dryers are rubbed on one side of a balloon, that side will not be able to pick up and hold an electric charge when rubbed against human hair, unlike the other, untreated side of the balloon.

   The law of charges can be demonstrated with two pieces of cellophane tape that are pressed down on a table top with tape extending over the edge. If they are rapidly pulled off the table, like sides of the tape will repel and unlike sides will attract one another. Students can explore real-world applications of static electricity such as electrostatic photocopying and the need for humidifiers in homes with hot air furnaces. The book *Safe and Simple Electrical Experiments* (1973) by Rudolf F. Graf contains 38 static electricity experiments. The NSTA Press book *Taking Charge: An Introduction to Electricity* (2001) by Larry Schafer is another valuable resource for grades 5–8. See also Activity #15 in this book.

Safety Note

The edges of aluminum soda cans can be sharp and cut the skin. Use caution in handling.

2. *Historical Connection.* Benjamin Franklin's international reputation (as the scientist who tamed lightning and introduced the idea of + and − "electric fluids") played a pivotal role in convincing the French to aid the American revolutionaries. The life of this fascinating polymath is explored in the 210 minute, PBS Home Video, *Benjamin Franklin* DVD (2002) #BENF601. *www.shoppbs.org/product/index.jsp?productId=1402894#Details*

Internet Connections

- American Educator, Ask a Cognitive Scientist column: *www.daniel-willingham.com.* (See (1)"Brain-based" learning: More fiction than fact: *www.aft.org/pubs-reports/american_educator/issues/fall2006/cogsci.htm* (2) Do visual, auditory, and kinesthetic (VAK) learners need VAK instruction?: *www.aft.org/pubs-reports/american_educator/issues/summer2005/cogsci.htm* (3) Inflexible knowledge: The first step to expertise: *www.aft.org/pubs-reports/american_educator/winter2002/CogSci.html*)

- Arbor Scientific's Cool Stuff Newsletter: *www.arborsci.com/CoolStuff/Archives3.aspx.* (See both electrostatics entries and newsletters #29 and #39 for great demonstrations.)

- Concord Consortium (free downloadable simulations): Electrostatics (see Polarization page): Molecular workbench software homepage: *http://mw.concord.org/modeler*

- Exploratorium Snackbook: Demonstrations involving triboelectricity: (1) *www.exploratorium.edu/snacks/charge_carry* (2) *www.exploratorium.edu/snacks/electroscope* (3) *www.exploratorium.edu/snacks/electrical_fleas*

- HowStuffWorks: Static electricity: *http://science.howstuffworks.com/vdg1.htm*

- PhET Interactive Simulations: Balloons and static electricity: *http://phet.colorado.edu/simulations/sims.php?sim=Balloons_and_Static_Electricity*

- Triboelectric Series: *www.school-for-champions.com/science/static_materials.htm.* (See also other pages on static electricity.)

- University of Iowa Physics and Astronomy Lecture Demonstrations (video clips of demonstrations): Electricity/Magnetism: *http://faraday.physics.uiowa.edu* (4 demos + triboelectric series)

- University of Virginia Physics Department: Static electricity (hands-on experiments and demos): (1) *http://galileo.phys.virginia. edu/outreach/8thGradeSOL/StaticFrm.htm* (2) *http://galileo.phys. virginia.edu/outreach/8thGradeSOL/NeonChargeFrm.htm* (3) *http:// galileo.phys.virginia.edu/outreach/8thGradeSOL/Salt&PepperFrm.htm* (4) *http://galileo.phys.virginia.edu/outreach/8thGradeSOL/ BalloonElectroscopeFrm.htm*

- University of Virginia Phun Physics Show: Electrostatics demonstrations: *http://phun.physics.virginia.edu/demos/electrostatics.html*

- Wake Forest University: Electricity videos: Electrostatics: *www.wfu. edu/physics/demolabs/demos/avimov/bychptr/chptr7_electricity.htm*

- Whelmers #31: Static charged 2 × 4s: *www.mcrel.org/whelmers/ whelm31.asp*

- Wikipedia: Electrostatics: *http://en.wikipedia.org/wiki/Static_ electricity* and Fabric softeners: *http://en.wikipedia.org/wiki/ Fabric_softener*

Answers to Questions in Procedure, steps #1–#4

1. This is a very open-ended question, but given universal natural laws that "govern" matter and energy, a wood board could quite arguably hold the keys to discovering many of these laws.

2. The balanced board could be moved by blowing on it. This would demonstrate that not a lot of force is needed given that the board is balanced on a fulcrum at its center of mass and that there is relatively little friction between the board and the watch glass. Power is a measure of the rate of energy use per unit of time. Although it is a glucose and oxygen "hog" in the body compared by mass to other organs, the brain operates 24/7 on slightly less than 25 W of power. The movement of the board has nothing to do with "brain power" and everything to do with charging by induction.

3. If you hold the balloon on the other side of the board, it will slow down, stop, and reverse direction (being attracted again to the negatively charged balloon).

4. Electric (and related magnetic) energy and forces exist throughout the universe and are ultimately related to the existence of charged subatomic particles. The law of charges will be explored in other activities, but objects with the same kind of charge will repel each other.

Activity 20

Needle Through the Balloon: Skewering Misconceptions

Expected Outcome

A long metal needle (or a bamboo skewer) is passed through a balloon without popping it. If it is pulled back out (or pushed all the way through), the balloon will slowly lose air.

Science Concepts

Plastic or latex balloons are made of long, cross-linked elastic polymers that form a flexible netlike matrix with variable size molecular "holes." According to the kinetic molecular theory, molecules in a gaseous state can diffuse out of "solid" balloons at a rate dependent on the relative size (and speed) of the molecules and the "holes" in the balloon. When a needle is carefully passed through the thicker ends of the balloon, the multilayered polymer seals itself around the needle until it is withdrawn.

Science Education Concepts

Prior experiences and conceptions that students have about a wide variety of science phenomena (e.g., sharp needles pop balloons) often form barriers to developing more scientifically valid understandings. Research suggests that core scientific ideas that are counterintuitive (e.g., atoms and the kinetic molecular theory) merit a multisensory instructional cycle (e.g., the 5E Teaching Cycle: Engage, Explore, Explain, Elaborate, Evaluate; see Appendix B for more information) that gives students multiple contexts, spread out over time, to develop understanding of the concept.

Discrepant events can be used to activate, assess, and challenge preexisting misconceptions and to identify conceptual "holes." The cognitive conflict or dissonance that such activities generate also creates an emotional and cognitive "hook" that catalyzes the cognitive construction associated with minds-on learning. This activity serves as a *visual participatory analogy* for the pedagogical challenge of strongly held misconceptions. Or, as Mark Twain put it, "It ain't what you don't know that gets you into trouble. It's what you know for sure that just ain't so."

Materials

(*Safety Precautions:* **Students who have allergic reactions to natural rubber latex should avoid contact with latex balloons, gloves, rubber bands, and so forth.** For more information on latex allergies, see *www.mayoclinic.com/health/latex-allergy/DS00621.*)
- 1, 18 in. metal needle
- 12 transparent balloons

Or go to
- Abracadabra Magic. *www.abra4magic.com.* 732-805-0200. #N009. $19.95.
- Creative Presentations. *www.presentationresources.net/magic_supplies. html.* #103-023. $9.95.

Alternatives:
- 12 in. or longer bamboo skewers (grocery store variety; lightly sanded to remove any loose splinters) or upholstery needles. You will also need 8–9 in. transparent round balloons.
- Vaseline or petroleum jelly rubbed lightly (or cooking oil or WD-40 sprayed) on the needle or skewers (will help lubricate and seal).
- Magnifying glasses can be used for misdirection (pretend to try to find the molecular holes in the balloons).
- *For Extensions:* Plastic baggies, pencils, soap bubbles, vanilla

Safety Notes

1. Use caution in working with needles and skewers, which can puncture the skin.
2. Students and teacher should wear safety glasses or goggles during this activity.

Points to Ponder

Conflict is the gadfly of thought. It stirs us to observation and memory. It instigates to invention. It shocks us out of sheep-like passivity, and sets us at noting and contriving…. Conflict is a sine qua non of reflection and ingenuity.

—John Dewey, American philosopher-educator (1859–1952)

Teaching methods based on research in cognitive science are the educational equivalents of polio vaccine and penicillin. Yet few outside the educational research community are aware of these breakthroughs or understand the research that makes them possible.

—John Bruer, American cognitive psychologist, quoted in the *American Educator* Summer 1993 (from his book, *The Mind's Journey from Novice to Expert;* see Internet Connections: John T. Bruer)

Procedure

The following demonstration can be done with teachers or students. It helps if you carry it out in the spirit of "science is magical!"

1. Begin the demonstration by blowing up several balloons to about one-half of their capacity. Tell the class that scientists have discovered that all matter is made up of constantly moving atoms (commonly bonded together as molecules) that are separated by variable-size, empty spaces that shift locations, called "holes." Ask volunteers to use magnifying lenses and bamboo skewers to find the "holes" in the inflated balloons you give them and subsequently pass the skewer through the holes. This challenge likely will result in several puzzled looks and loud balloon pops.

2. Announce: "Advanced training in chemistry enables a master teacher to find and align the skewer to the molecular holes." If available, dramatically pull out a long magic-store needle and use a gentle twisting motion to pass it through a balloon at the two thick ends (i.e., near and directly opposite the tie-off point). The balloon will pop if you insert it anywhere else on the balloon (where the balloon membrane is thinner). You may wish to stop and enjoy your success when you have successively gone through one end of the balloon. If the partially skewered balloon is held over an overhead projector (used as a light table), the portion of the needle inside the balloon will be made more visible. If the needle is pulled back out at this point, the balloon will lose air slowly as the hole only partially reseals itself as air is released.

   Alternatively, tempt fate and proceed to slowly insert the needle all the way through the opposite end of the balloon. The skewered balloon will lose air quite slowly if the needle is left in place. Even an experienced demonstrator will sometimes pop a balloon. Be playful about such "miss-takes"; after all, finding molecular holes is not easy!

3. (For the answers to questions in this step, see p. 219.) If time permits, allow all the learners to try to replicate your science magic trick using bamboo skewers. However, first ask: What was different about my approach versus that of your peers who tried to do

214

it before I did? If they think the needle is the trick, repeat your demonstration using a bamboo skewer but ask them to more carefully observe your actions this time. Ask: What questions would you like to explore related to this phenomenon? How would a chemist explain this seemingly discrepant event (i.e., a needle that doesn't pop a balloon)? Are solids really solid at the molecular level? Does a scientific explanation preclude a sense of mystery and awe?

4. If an additional discrepancy is desired, secretly place a small, inconspicuous piece of transparent tape on the side of a balloon without showing the class. You will be able to insert the needle through the side of this "rigged" balloon, a "trick" that is otherwise not possible. If you proceed to push it all the way through the other side, the balloon will pop (unless that side is also rigged). This principle is used in some types of blowout-proof, self-sealing automotive tires.

Debriefing

When Working With Teachers

This demonstration serves both as a visual participatory analogy for and specific example of the role of multisensory experiences (e.g., sight, sound, and touch) in assessing and challenging misconceptions in learning science. Some ideas that earlier scientists believed to be true—and that many students today believe—need to be modified or replaced. Historically important, but now outdated, scientific theories and current day student misconceptions commonly are flawed in one of two ways:

1. They need to be modified to account for a broader range of observed phenomena. For example, the model of atoms as solid billiard balls can still explain gas laws and the conservation of matter, but we use a solar system model with a positively charged nucleus and negatively charged electrons to account for electricity. Even more complex models are needed to account for electromagnetic interactions with matter. Thus, although earlier conceptions are technically incorrect, they are still useful as mental models to explain some simpler phenomena.

2. Some earlier scientific theories (and student conceptions) are so limited in their range of valid application that they need to be "popped" and completely replaced by more powerful explanatory theories. Examples of this from the history of science include the idea that gases are "no thing" and therefore do not really matter and that solids are truly solid (versus particulate). Interestingly, many misconceptions that students bring to class reflect outdated theories of science that have long since been completely replaced in scientific work.

Discuss the quotes by Dewey and Bruer relative to the role of intentionally creating cognitive conflict as a research-informed pedagogical technique. An example of an article that will challenge teachers' commonsense understanding of the brain is the online cover story in the Spring 2009 issue of *American Educator*, "Why Don't Students Like School? Because the Mind Is Not Designed for Thinking" (by Daniel T. Willingham, the author of the book by the same name). You might also view the Teacher-Tube video "Brain-Based Education: Fad or Breakthrough?"

When Working With Students

This activity is a good Engage-phase activity to generate student questions about the atomic nature of matter. Let students enjoy and even be intellectually irritated by the mystery and magic. Remember, an oyster's response to an irritating substance is to produce, over time, a pearl of great value.

Extensions

1. *Variation #1a: You're All Wet.* Sharpened pencils can be quickly poked straight into a partially water-filled polyethylene bag ("baggie") without loss of water if the pencils are left in place. This is more dramatic if done over a sitting volunteer's head. Polyethylene is a thermoplastic (i.e., melts with heat) that will shrink around a hole to partially self-seal itself as long as the pencil is not removed.

Variation #1b: Milk Bottle Magic. White, semi-opaque milk containers are made of the same polymer as baggies. Direct a heat gun (as used for stripping off paint) in a circular motion at one spot on an empty, rinsed milk jug and watch it slowly turn into a clear, colorless plastic. If you quickly blow on the mouth of the jug, you can blow a baggie out the side of the bottle! Membranes are magical!

2. *Variation #2: Biological Cell Membrane Analogy.* Consider how spherical balloons and soap bubbles are like and unlike the fluid mosaic model of cell membranes. All three systems are alike in the following ways: they are 3D, roughly spherical, semipermeable, self-sealing, multilayered, thin, and membrane-bound. Also, they all contain different internal and external environments that exchange materials partly driven by diffusion. (See Internet Connections: Access Excellence and Nanopedia websites.) The cell needs to excrete waste and take in oxygen and glucose to maintain a state of homeostasis amidst constant, 24/7 biochemical reactions. The cell membranes consist of a phospholipid bi-layer that utilizes active as well as passive transport to maintain the cell membranes' internal environments in a state of dynamic equilibrium. A soap bubble is a bi-layer (but with water between the two layers of soap). Soap bubbles are also self-sealing if a wet object (e.g., a finger or straw) is inserted through the bi-layer. Soap bubbles can be blown within soap bubbles to model eukaryotic cells with their internal, membrane-bound organelles. Also consider how viruses invade cells and initially insert their genetic material into the cell without the cell popping. Regardless of the grade level, consider using 3D macroscopic models in combination with microscopic explorations.

3. *Variation #3: Molecules Matter and Odiferous Orbs.* (For answers to the questions in this step, see p. 219.) Food extracts such as vanilla and peppermint can be "observed" to diffuse through the "molecular holes" in pre-stretched, breath-inflated balloons into which an eyedropper of extract has been inserted before the balloons are blown up and tied off. If several of these balloons are passed around the room and waved back and forth a few times, the smell will soon permeate the room; the liquid will be observed to "disappear" from the inside of the balloon; and the balloons will feel slightly cooler to the touch (due to the heat absorbed from the balloon in the liquid-to-gas phase change). Have students use the idea of molecular holes and moving molecules to explain balloon basics by answering the following questions:

- Why do human breath–filled balloons lose "air" or "leak" over time?
- Why does helium escape from standard latex balloons much faster than from human breath?
- Why do mylar balloons hold their helium so much longer than latex ones?

Internet Connections

- Access Excellence: Using (soap) bubbles to explore (cell) membranes: *www.accessexcellence.org/AE/AEC/AEF/1995/wardell_membranes.php*

- American Educator (see also *www.danielwillingham.com* for other articles and videos): Critical thinking: Why is it so hard to teach? (feature article): *www.aft.org/pubs-reports/american_educator/issues/summer07/index.htm* and Why don't students like school? Because the mind is not designed for thinking (feature article): *www.aft.org/pubs-reports/american_educator/issues/spring2009/index.htm*

- John T. Bruer. Publications and presentations on cognitive psychology. Free downloads. *www.jsmf.org/about/bruer-publications.htm*

- David Katz's Needle Through Balloon (explanation): *www.chymist.com/Balloons.pdf*

- Nanopedia: Soap bubbles as nanoscience (with embedded link to cell membranes): *http://nanopedia.case.edu/NWPage.php?page=soap.bubbles*

- Science Hobbyist: Balloon demonstrations: *http://amasci.com/amateur/balloon.html*

- Wikipedia: Membranes (biological and artificial): *http://en.wikipedia.org/wiki/Membranes* and *http://en.wikipedia.org/wiki/Run-flat_tire*: Section 1.2 Self-sealing, run-flat tire

Answers to Questions

Procedure, step #3

The key is to insert the skewer slowly (with a little twisting and/or lubrication) *through the two ends* (versus elsewhere on the balloon where the polymer membrane is thinner). "Holes," or empty spaces, exist between (and also within) atoms and molecules, but atomic and molecular motions and the thickness of most materials typically prevent us from being aware of them. (Later in a unit on this topic, you might clarify that the needle has a much larger diameter than any molecular hole and that the trick is due to the self-sealing nature of the multilayered, thick, elastic polymer [i.e., the polymers act somewhat like wet spaghetti in wrapping themselves around the hole]. Early in the unit, avoid the urge to tell the "(w)hole story." True science always creates at least two questions for every one it answers.)

Extension Variation #3: Molecules Matter and Odiferous Orbs

Any gas-filled rubber or latex balloon will lose some of the internal pressurized gas as the molecules effuse through the molecular holes in the membrane. Smaller gas molecules (such as hydrogen and helium) will leak out faster than the nitrogen, oxygen, carbon dioxide, and water vapor found in human breath for two reasons. First, given their smaller relative size, they can more easily find their way through multiple layers of latex. Second, given that temperature is a measure of average kinetic molecular energy and that $KE = \frac{1}{2} mv^2$, less massive molecules at a given temperature, on average, move faster than larger molecules at the same temperature. Also, if two balloons were filled with the same gas but at two different temperatures, the warmer, faster moving molecules would leak out faster. The metallic coating on mylar balloons has aluminum atoms packed tightly together to help fill in many of the holes in the polymer; therefore they are much more effective at retaining trapped gases. One can also buy a liquid coating to put inside latex balloons that helps somewhat to retain trapped gases.

Activity 21

Happy and Sad Bouncing Balls: Student Diversity Matters

HAPPY BALL

SAD BALL

Expected Outcome

Two seemingly identical black rubber balls behave quite differently. One ball bounces (when dropped) and rolls down ramps very well (the "happy" ball); the other ball (the "sad" ball) barely bounces at all and is slightly slower rolling down a ramp and across the floor (i.e., it will "lose" in a fair competition against the happy ball). Also, an inverted half racquetball will jump up high above its release point.

Science Concepts

At lower grade levels, the two balls can be used to introduce the National Science Education Standards' unifying concepts and processes: *change, constancy, and measurement.* At higher grades, *evidence, models, and explanation; form and function;* and (in life science classrooms) *evolution* can be explored in more depth (NRC 1996, p. 104).

On the topic of evolution, consider the following ideas. The physical properties of substances depend on their chemical composition (as well as on the ambient temperature and pressure). The composition of various synthetic rubber polymer formulations is intentionally designed by humans (similar to the artificial selection of animal breeding) to meet particular environmental conditions and user needs (e.g., automobile tires and aircraft tires are designed for different temperature ranges and for desired degrees of friction). An analogy can be drawn to the science concept of natural selection whereby the biological fitness of an organism for a given environment (form to function alignment) drives evolutionary changes. More broadly, this activity models the nature of science in cases where scientists probe beneath surface-level appearances of sameness to discover uniqueness.

These two balls have the same color and volume (but slightly different mass and density due to their different chemical formulations). When dropped, the "happy" one rebounds like a Super Ball (i.e., has good elasticity and reaches ~90% of its release height) and the other, "sad" ball hardly bounces at all (i.e., it is quite inelastic). Similarly, the sad ball is slower when going down a ramp due to its greater rolling friction (i.e., more of its gravitational potential energy is converted into random molecular motion and "waste" heat in both cases). However, one type of rubber cannot be judged to be universally superior to the other; the characteristics of rubber depend on the properties desired for a particular environmental context and use. Also, a racquetball cut in half (or a commercial Dropper Popper toy) can store energy if it is inverted. When it is dropped concave side down, it will rebound above the level it was dropped due to the potential energy stored in its distorted shape (thus qualitatively demonstrating the conservation of energy).

222

Science Education Concepts

This activity can be used as a *visual participatory analogy* for student cognitive and emotional diversity and the need for scaffolded curriculum-instruction-assessment that meets students "where they are" and takes into account their varying multiple intelligences, abilities, and motivations. Simply put, the learning environment must be designed to help all students achieve their full potentials rather than merely serve the few students who can adapt to a one-size-fits-all approach. Emotional engagement, connections, and relevance are essential elements of effective learner-centered teaching.

Materials

- Happy (polyNeoprene) and Unhappy/Sad (Norsorex/polynorbornene) Ball Set:
  - Arbor Scientific. *http://arborsci.com.* 800-367-6695. #P6-1000. $3.95/two 1 in. balls and instructions
  - Edmund Scientific. *http://scientificsonline.com.* 800-728-6999. Two different-size sets available: #3039129, ¾ in. diameter/$5.95 and #3082296, 1.5 in. diameter/$8.95 (with fact sheets)
  - Educational Innovations. *www.teachersource.com.* 888-912-7474. #SS-3. $7.75/two 1.5 in. balls.
- *Optional:* Animated Happy/Young–Sad/Old 180 degree optical illusion (see Internet Connections: Illusion-Optical.com)
- *Optional Extension #1:* 2, 1 m lengths of corner molding or PVC tubing to use as ramps
- *Optional Extension #2:* Arbor Scientific sells the Dropper Popper (#P6-6075. $2.75), as do toy stores. Better yet, you can easily make your own.

Safety Note

Students and teacher should wear safety glasses or goggles during this activity.

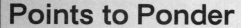

Points to Ponder

An exaggerated competitive attitude is inculcated into the student, who is trained to worship acquisitive success as a preparation for his future career.

—Albert Einstein, German-American physicist (1879–1955)

One of the most important remedies for combating the illusion of understanding and the persistence of misconceptions is to support learners in the active, collaborative, reflective reexamination of ideas in a social context. Learning is least useful when it is private and hidden; it is most powerful when it is public and communal. Learning flourishes when we take what we think we know and offer it as community property among fellow learners so that it can be tested, examined, challenged, and improved before we internalize it.

—Lee Shulman, American educator and researcher (1930–) in "Taking Learning Seriously," *Change: The Magazine of Higher Learning* 31 (4), 1999

Procedure

When Working With Teachers

Use the two balls as a visual participatory analogy. (See answers to the questions in steps #1 and #2 on pp. 230–231.)

1. Hold up the two balls for the teachers to see and ask: Are our students more or less all the same? What kinds of diversity are present in our classes? Even in a classroom that has limited ethnic, racial, religious, and socioeconomic diversity (i.e., all the students "look the same," like these two balls), is it safe to assume that students have the same multiple intelligences profiles, cognitive abilities, prior knowledge, and motivations for learning? How can we assess whether these two balls (as analog models for our students) are the same or different?

Elicit the teachers' suggestions and then drop the balls at the same time from the same height and note the discrepant outcomes. If time permits and you have a sufficient number of sets of balls, teacher teams can pursue independent or guided investigations of the balls' properties (see When Working With Students p. 226). Minimally, you may want to set up a quick demonstration to see how the balls fare when rolling downward in a "ramp race."(*Note:* Galileo designed ramp races to "slow down" the effect of gravity so he could better study it with the primitive time-keeping technology of his era. He discovered the idea of gravitational acceleration as a result. Similarly, as teachers, we need to discover a variety of means to "accelerate" student learning. This may involve scaffolding—even if scaffolding seems, in the short term, to slow down the rate of instruction.)

2. As the teachers note the dramatic difference in the balls' rebounds (and a less dramatic, but noticeable difference in their race times), initiate live or online, whole- or small-group, short-term or year-long discussions related to analogical questions such as the following:

   a. Does your current curriculum-instruction-assessment plan create connections and relevance for all (or only some) of your students? How much of a sad or slow ball's ("weak" students') inabilities and disabilities might be due to curriculum-instruction-assessment mismatches?

   b. Could some of the happy, super balls' ("good" students') success be due to a narrow band of multiple intelligences that are being assessed or to the assessment conditions? Test how a happy ball rebounds off a soft, cushy material such as foam.

   c. Are high-achieving students being challenged to stretch and grow in new directions or is schooling just reinforcing what they naturally do well?

   d. Regardless of where the "blame" lies or how it is partitioned (i.e., among the individual student, the home, the school, the community, or the broader society), how can teachers reverse the pattern of past failures of sad ball–type students?

   e. How can student diversity be used to create a rich cooperative (versus an unduly competitive) learning environment?

f. How can students' inherent potential energy to learn be released so that all students can reach their true potentials (versus being limited by artificially low expectations)?

g. How can cooperative learning activities use diverse student strengths to help overcome individual weaknesses?

h. How does the decreasing height of each subsequent bounce of the happy ball model student retention of memorized information over time? Would meaningful learning show this same kind of "decay rate"?

Optional: Use the 180 degree happy/sad optical illusion to stress how effective teachers can reverse the negative attitudes that some students have about science. Ask: How can teachers ensure that students come to class with anticipation and leave with regret (rather than the reverse)?

When Working With Students

This activity can be used (either as a demonstration or as hands-on explorations if multiple sets of balls are available) in a unit on energy that includes topics such as conservation of energy, gravitational potential and kinetic energy, and friction. (See answers to questions in steps #1–#3 on pp. 231–232.)

1. Show the students the two balls and ask: Is it easier to "prove" that these two balls are identical or that they are different? What measurements or tests provide adequate evidence? You could also initiate this inquiry as a sleight-of-hand magic trick at step #2 (below) by acting as if you have only one ball but then secretly substituting the sad for the happy ball (after the first "successful drop") to produce a discrepant-event result that demands explanation. This "trick" can be used to highlight the difference between an observation and an inference.

2. Drop the happy ball and ask students to note how high it rebounds relative to the point of release. Ask questions such as these:

   a. What kind of energy does a raised, but stationary, ball contain? How could we make it bounce higher or lower?

   b. Can a dropped (not thrown) ball rebound to a higher level than its point of release? Why or why not?

   c. Why do we expect dropped balls to rebound to a height less than their point of release and rise to a lower height with each progressive bounce? Where is the original energy going?

3. Substitute the sad ball and repeat the drop from the same height. As students note the almost total lack of rebound or bounce-back, ask them to think about their answers to question 2c and to come up with tests they would like to run on the sad ball to determine the following:

   a. Why does this ball behave so differently from the happy ball?

   b. Would the two balls roll down a ramp differently? If so, predict-observe-explain how they will vary based on their bouncing behavior?

   c. How will higher or lower temperatures affect their properties? (See Internet Connections: University of Virginia Physics Department website.)

   d. Would throwing (versus dropping) the ball make a difference? Try it!

   e. Brainstorm a list of practical, real-world applications where the rubber formulation in the sad ball would provide an advantage over that of the happy ball.

Debriefing

When Working With Teachers

Use this activity to raise questions about student diversity and teachers' ethical responsibilities to care for and help all students learn, both despite past failures and beyond any past successes. Use the Extension Demonstration #2 and the Einstein and Shulman quotes to emphasize the interactive and experimental nature of teaching and learning and the need for emotional engagement, connections, and relevance in the classroom. Recall this truism: "Students won't care how much we know until they know how much we care." Teachers could individually or collectively explore both the pedagogical implications and criticisms of various theories of intelligence, differentiated instruction, and other "brain-based" education ideas (see Internet Connections).

When Working With Students

Beyond energy-related concepts, the happy/sad balls serve as a nice "black box" system to explore and discuss how chemists and chemical engineers meet specific societal needs and technical specifications by building "designer molecules." Also, in life science or biology classes, the two balls can be used as an analogical model for discussing how evolution leads to a "fit" between organism and environment. Also, note that what is considered "fittest" depends on the particular environmental conditions present at a given point in time and space. (For example, mammals were "minor league players" during the long Age of Dinosaurs until dramatic astronomical events altered the Earth's global environment and changed the playing field, making mammals the "winners" and reptilian dinosaurs the "losers.")

To identify one ball as universally better is not appropriate, nor can one organism be said to be better adapted independently of its environment. High friction, low bounce polymers (or balls) have important applications. Similarly, organisms that lack eyes or other attributes that humans might consider essential may be better adapted to certain environmental conditions (i.e., the cost/benefit ratio of "investing" in eyes varies with the organism's environment and lifestyle).

Extensions

1. *Momentum Matters.* The two balls can be used to explore conservation of momentum when one ball is placed on top of the other and they are dropped together in line. The classic version of this demonstration is done with two balls of different masses, with the smaller, lighter one on top (e.g., a golf ball and a Ping-Pong ball or a basketball and tennis ball; see Internet Connections: University of Virginia Physics Show). Compare how two happy balls would respond versus two sad balls versus a happy and a sad ball together (should it matter which is on top?).

2. *Dropper Popper (High Hopper).* This toy is actually the ultimate Super Ball in that if it is inverted or turned inside out (storing your muscular mechanical energy in its deformed shape) and dropped concave side down, it will jump back to a level much higher than its release point. The cheaper dollar store and toy store varieties sometimes don't hold their shape well enough to be dropped.

They will however leap up if they are placed in their inverted shape directly on the floor.

An even better-performing, homemade version of this toy can be made by cutting an old racquetball along its seam and then trimming it down until it has a 5.5 cm external diameter (see Internet Connections: Becker and University of Virginia Physics Show websites). This size will hold its inverted shape until it is dropped. If the half ball doesn't jump when dropped, you need to trim it down more; if it won't hold its inverted shape, you need to throw it away and start over! Cut tennis balls will also work, but they do not rebound quite as well. These half balls can be used as a visual analogy to model the potential of even seemingly "flawed" students to achieve well beyond the self-limiting, deficit model of low school expectations.

3. *The Incredible Flubber.* Clips from the original or more recent version of the Disney movie *Flubber* (starring Fred McMurray or Robin Williams, respectively) provide a playful look at the properties and potential uses of polymers.

4. Activity #2 in this book, especially the Extension with the Ernest the Balancing Bear toy, can also be used to model the idea of emotional engagement and connections as essential attributes of effective teaching.

Internet Connections

- Becker Demos (chemistry movies): Captivating activation (uses a half racquetball as analogy for chemical activation energy for an exergonic reaction): *http://chemmovies.unl.edu/chemistry/beckerdemos/BD054.html*

- CAST: Differentiated instruction: *www.cast.org/publications/ncac/ncac_diffinstruc.html*

- Concept to Classroom: Multiple intelligences tutorial: *www.thirteen.org/edonline/concept2class/mi/index.html*

- Human Intelligence: New and emerging theories of intelligence (synopses of Gardner's theory of multiple intelligences, Sternberg's theories, and Goleman's theory of emotional intelligence): *www.indiana.edu/~intell/emerging.shtml#intro*

- Illusion-Optical.com: Happy (young princess)-Sad (old woman) 180 degree animation: *www.illusion-optical.com/Optical-Illusions/HappySad.php*

- Reframing the Mind: A critique of the theory of multiple intelligences (Daniel Willingham): *http://educationnext.org/reframing-the-mind.* See also other articles and critiques of brain-based education fads by the same author: *www.danielwillingham.com*

- University of Iowa Physics and Astronomy Lecture Demonstrations (video clips): Mechanics coefficient of restitution: *http://faraday.physics.uiowa.edu* (two demonstrations with balls)

- University of Virginia Physics Department: HOEs: Energy of a bouncing ball and temperature effects: *http://galileo.phys.virginia.edu/outreach/8thGradeSOL/EnergyBallFrm.htm* and *http://galileo.phys.virginia.edu/outreach/8thGradeSOL/EffectofTemperatureFrm.htm*

- University of Virginia Physics Show: *http://phun.physics.virginia.edu/demos/hopper.html* and *http://phun.physics.virginia.edu/demos/double.html* (double ball bounce)

Answers to Questions in Procedure, When Working With Teachers, steps #1 and #2

1. Each student has his or her own unique multiple intelligence profile (and range of abilities and interests) based on the interaction of genetic, environmental, self-directed choice, and effort factors. Each student also brings unique prior experiences and conceptual networks of understandings (and misunderstandings) to any new experience. Therefore, each student uniquely and selectively perceives, conceives, retains, and retrieves his or her own individual interpretations of a given experience. Cognitive and emotional diversity among individuals is much greater than the more easily observable differences between the two sexes, between ethnic and racial groups, and between socioeconomic groups.

2. These questions are provided to elicit reflective conversation on teaching practices and school policies that may narrowly define intelligence or "smarts." In this step in the Procedure section, the author's intention is not to move teachers to the "right" answers but rather to have them consider focusing more on how each and every student is smart rather than on how smart a student is. Effective teachers build on unique student strengths and help students overcome or work around their weaknesses. Meaningful learning does not show the rapid retention drop or decay that occurs with mindless memorized information; it tends to be both truly memorable (lasting over time) and flexible (transferable to different contexts).

Answers to Questions in Procedure, When Working With Students, steps #1–#3

1. One *falsification* test could prove if the two balls are different, whereas, theoretically, an infinite number of tests would be needed to prove them identical. However, for practical purposes, scientists would have confidence with a more limited number of *verification* tests. It is likely that students will suggest that you simply drop and/or roll the balls to see if they are different, but mass, volume, and density tests and temperature checks may also be suggested.

2a. A raised stationary ball contains gravitational potential energy that is directly proportional to its height. The potential energy can be raised or lowered with the use of muscular, mechanical energy.

2b. No. If a dropped ball rebounded to a higher level than its point of release this would contradict the law of conservation of energy. The ball cannot release more kinetic energy than the gravitational potential energy it had due to its initial height.

2c. With each successive bounce, some of the energy is converted into random molecular motion associated with frictional heat and sound. (Do not give students all this information now.)

3a. Both balls are black and are the same volume and temperature. The small difference in mass leads to density differences that affect the balls' ability to float or sink in water. Unless one of the two identical-volume balls has a hollow interior, different densities suggest that the two look-alike balls are made of different materials because density is an intrinsic property of matter.

3b. The low-bouncing sad ball also has increased rolling friction and will lose a fair race with the happy ball.

3c. Interestingly, if the two balls are placed in hot water that is brought to near boiling (the happy ball will float around 85°C) and then are removed, their properties reverse somewhat (i.e., the sad ball will bounce better than it did before and the happy ball less well; see Internet Connections: University of Virginia Physics Department). Alternatively, if they are both placed in the freezer, the bouncing performance of *both* balls is reduced.

3d. The sad ball does not rebound much better when it is thrown than when it is dropped.

3e. The nonbouncing property of the sad ball can be used as a component of automobile tires to absorb some of the energy from bumps and potholes in the road without causing the car to bounce up and down excessively.

Electrical Circuits: Promoting Learning Communities

Expected Outcome

Direct current (DC) electricity flows through a closed circuit of people, and a battery-powered ball lights up.

Science Concepts

The Energy Ball (or UFO Ball) is a 1.5 in. Ping-Pong ball look-alike, battery-powered ball that produces and converts a small current into light and sound energy. It contains two small, external metal contacts that when touched simultaneously (by a single person or by two or more people forming an unbroken chain) complete an electrical circuit and cause the ball to flash and emit sound. Although the resistance of *dry* skin is very high (~500,000 Ω), a tiny, yet detectable current (about a millionth of an amp) will flow through the human body with the application of only a few volts. Topics that can be explored include electrical conductivity, series (and parallel) and closed (and open) electrical circuits, and energy conversion.

Science Education Concepts

If students are to invest the cognitive effort required for learning, they must experience emotional engagement, connections, and relevance. This idea is captured in the teacher adage that students won't care how much you know until they know how much you care (both about them as individuals and your discipline). Given the evolutionary history of the different layers of our brain (i.e., the limbic system predates the cerebral cortex), it can be said that humans are more feeling-creatures-that-think than we are thinking-creatures-that-feel. Given our highly developed intraspecies communication skills, we are a highly social species that can learn through structured collaborative interactions among peers. Learning communities in which students cooperatively help each other learn are more effective than classes that over-emphasize competition for the teacher's "limited supplies" of attention and external praise and rewards. Or, to use a line from the opening song for the TV show *Cheers* (1982–1993), the optimal learning environment is one "where everybody knows your name, and they're always glad you came."

In a similar fashion, teacher-to-teacher collaborations promote the dissemination of best practices. In contrast to scientists, science teachers are often isolated from their peers whether the peers teach down the hall, in an adjacent school, or across the country. This lack of symbiotic, synergistic networking and sharing of successes limits the advance of science education as a field. The discrepant-event demonstration serves as a *visual participatory analogy* for building connections

234

within and between classrooms, while simultaneously serving as an engaging hands-on and minds-on way to get students thinking about electrical circuits. It can also be used to teach teachers the importance of helping students make connections between the beginning of a lesson (i.e., set induction or cue set) and an explicit closure that helps the lesson come full circle to achieve the target objectives.

Materials

- 1 or more Energy Balls (also called UFO Balls) (or similar electrical circuit novelty toys):
  - Arbor Scientific. *http://arborsci.com*. 800-367-6695. #P6-2300. $3.50
  - Edmund Scientific. *http://scientificsonline.com*. 800-728-6999. #3052930. $5.95
  - Educational Innovations. *www.teachersource.com*. 888-912-7474. #SS-30. $3.25
- Similar products disguised as "chirping" bunny rabbits or ducks are available in drugstores and dollar stores at Easter time. An assortment of metal, plastic, and wood utensils and an ordinary Ping-Pong ball are also useful to compare and contrast conductors and insulators.
- *Optional Music:* "Circle of Life" (from *The Lion King*) or "Will the Circle Be Unbroken"

Points to Ponder

If I have seen a little further it is by standing on the shoulders of Giants.

—Isaac Newton, English scientist and mathematician (1642–1727)

What a person thinks on his own, without being stimulated by the thoughts and experiences of other people, is even in the best case rather paltry and monotonous.

—Albert Einstein, German-born American physicist (1879–1955)

Procedure

(Answers to questions in steps #2 and #3a are found on p. 240.)

1a. *When Working With Teachers:* Introduce the science education activity by noting that teaching is the science and art of playfully engaging learners to motivate them to undertake the hard work of learning so that each one can light up to his or her potential or sing his or her own unique song.

1b. *When Working With Students:* Use this science activity as an Engage-phase discrepant event to raise questions about electrical circuits. Skip to Explore activities in this book, such as Activity #9, before going to steps #3b–#4 as part of an Explain and/or Elaborate phase of a 5E Teaching Cycle.

2. Invite a learner up to the front of the group to role model being a teacher of a "typical" student (i.e., the Ping-Pong ball). Put the ordinary Ping-Pong ball ("student") in the open hand of the volunteer ("teacher") and ask if he or she can make this model student "light up and sing." When the volunteer is unable to do so, ask if it is the "fault" of the teacher or the student? Immediately follow this question by doing the same experiment in your hand using the Energy Ball (or similar toy) that will be observed to light up and/or make a sound. Give this ball to your volunteer to see if it will perform for him or her as well.

3a. Allow the volunteer to discover how to use the two metal contact points to complete the circuit and then ask, What forms of energy and scientific principles are involved in T.O.Y.S. such as this one? Consider sharing one of the following playful phrases that the acronym T.O.Y.S. can stand for: **T**errific **O**bservations and **Y**earnings for **S**cience and **T**est **O**f **Y**our **S**cience knowledge.

3b. Ask the learners to suggest other related experiments. As time permits, perform one or more of these experiments, but be sure to raise the question as to whether electrical energy can be conducted through a multi-human closed chain or circle. This question leads to the next visual participatory analogy.

4. Suggest to the learners that the effectiveness of any class depends in part on the connections and collaborations between teachers

and students and among students. To visually model this cooperative learning community, ask the learners to form a circle with 3 to 10 learners holding hands. Two people are not in the circle—they will touch the two metal contacts on the Energy Ball (without directly touching each other). The ball will light up and sing as before. If any member of the group breaks the chain, the ball will stop working. Metal spoons held between two learners in the circle will allow the energy to flow unbroken by completing the circuit. Wood or plastics, being nonconductors (or insulators), will not. Consider using one of the optional songs in the Materials list as a low-volume, playful audio background during this step of the activity.

Debriefing

This toy and activity is a visual participatory analogy for the nature of science as a community-based, constantly evolving enterprise. Many students and teachers have the mistaken image of the great scientists as being "solo artists." Use the quotes from Newton and Einstein to challenge this misconception. Newton's work (i.e., "standing on the shoulders of giants") built on that of Galileo who died the same year that Newton was born. Galileo credited the ancient Greek "scientist" Archimedes as being his inspiration. For older students, a brief walk through the history of science can be used to emphasize the community nature of science.

Modern science arose after the development of the printing press (by Johann Gutenberg in 1454 in Germany) allowed for the rapid dissemination and exchange of knowledge and the subsequent European explorations into the "New World." The earliest scientific societies (i.e., Italy's Academia Secretorum Naturae in 1560; England's Royal Society for the Improvement of Natural Knowledge in 1663 [built on the previous 1645 Invisible College for the Promoting of Physico-Mathematical Experimental Learning]; and France's Académie Royale des Sciences in 1666) were designed to encourage synergistic interactions among members and to facilitate intra- and international communications of scientific results (even during times of war between countries!). Although Leonardo da Vinci (1452–1519) created artistic masterpieces that continue to enchant the world, his failure to publish and share his scientific and engineering studies meant he had a limited impact in these domains.

Unlike a piece of art produced by a master artist, the work of individual scientists is always improved, refined, and in some cases completely replaced by other scientists. In a similar fashion, collaborations among teachers, classrooms, schools, and school districts could greatly accelerate the discovery and dissemination of pedagogical knowledge and best practices. Teachers can explore the theory and practice of cooperative (or collaborative) learning at the Cooperative Learning Center, Doing CL, and the Success for All websites (see Internet Connections) as a topic for individual study or study groups.

Extensions

1. *People and Parallel Circuits and Other Playful Investigations.* Test to see if there is a limit to the number of people (or the length of the series circuit) that will allow the Energy Ball to work. Determine whether the time lag between forming the circle and the ball's activation increases as the people circuit gets longer. Also, explore the effect of placing two UFO Balls in the series circuit. Can both balls be simultaneously activated? Is there a polarity effect? If both balls are activated and you rotate one of the balls 180 degrees, will both balls still emit sound and light? That is, does the order of the two electrodes matter when two balls are part of the same circuit?

    Try to make a parallel circuit using the Energy Ball and see what happens when one of the two pathways is broken [the ball still lights up]. Challenge students to find and explore the science behind other light-up toys from novelty shops and online sources (e.g., yo-yos, spinning tops, and magnetic spinning wheels). Many of these toys make use of centrifugal switches to complete the circuit that is powered by a battery.

2. *UFO Balls: A Black Box System.* High school physics students may wish to take apart the UFO Ball and explore its inner workings. Clearly this system is more complicated than Activity #9 ("Batteries and Bulbs") because human skin will not serve as a conductor to complete that series circuit. Arbor Scientific's website states that the ball "utilizes a field transistor." Students can explore the underlying science via the internet, library, and further experimentation.

Internet Connections

- All About Electricity: Online text: DC circuits: *www.allabout circuits.com/vol_1/index.html*

- Arbor Scientific's Cool Stuff Newsletter: *www.arborsci.com/ CoolStuff/Archives3.aspx* (See Electricity: Energy ball, people circuits, and other demonstrations.)

- Cooperative Learning Center at the University of Minnesota (Roger Johnson and David Johnson): *www.co-operation.org*

- Doing CL (Collaborative learning theory, instructional strategies, bibliography, etc.): *www.wcer.wisc.edu/archive/CL1/CL/doingcl/ DCL1.asp*

- Flash Animations for Physics: Electricity and magnetism: Compare a DC circuit to flow of water: *http://faraday.physics.utoronto.ca/GeneralInterest/Harrison/Flash/#sound_waves*

- Learning and Understanding Key Concepts of Electricity: Research article on student conceptions: *www.physics.ohio-state.edu/~jossem/ ICPE/C2.html*

- Lemelson Center for the Study of Invention and Innovation/ Edison Invents: Make a light bulb: *http://invention.smithsonian.org/ centerpieces/edison/000_lightbulb_01.asp*

- PhET Interactive Simulations: DC circuit construction kit: *http:// phet.colorado.edu/simulations/sims.php?sim=Circuit_Construction_ Kit_DC_Only*

- A Science Odyssey: Simple circuits, electromagnets and Morse code: *www.pbs.org/wgbh/aso/resources/campcurr/ telecommunication.html*

- Success for All Foundation (Robert Slavin and cooperative learning): *www.successforall.org/about/index.htm*

- Surfing Scientist: Nine pages of lessons (e.g., simple circuits and conductivity tester): *www.abc.net.au/science/surfingscientist/pdf/ lesson_plan11.pdf*

- University of Virginia Physics Department: Conductors and insulators (how a light bulb works) + Series and parallel circuits hands-on explorations: *http://galileo.phys.virginia.edu/outreach/8thGradeSOL/ Conductors&InsulatorsFrm.htm* and

*http://galileo.phys.virginia.edu/outreach/8thGradeSOL/Series
ParallelFrm.htm*

• Virtual Circuit Simulator/Lab: *http://jersey.uoregon.edu/
vlab/Voltage*

Answers to Questions in Procedure, steps #2 and #3a

2. The failure of the ball to light up and sing might be related to problems with either the "teacher" and/or the "student." Teaching effectiveness varies along a continuum and although students are never "empty balls," they may lack the necessary prerequisites to be able to perform a particular kind of on-demand test or task. In this case, the Ping-Pong ball lacks the internal circuit and light and sound generator that are present in the Energy Ball.

3a. The Energy Ball can be used to explore closed versus open electrical circuits, conductivity, and the energy conversions from chemical potential energy of a battery → electrical energy → light and sound.

Activity 23

Eddy Currents: Learning Takes Time

Expected Outcome

A metal slug dropped into a copper tube falls under the pull of gravity and drops out at the bottom fairly quickly, as expected. When a second, apparently identical (but actually a strongly magnetic) slug is dropped into the tube, it falls quite slowly. If one slug is secretly held in the instructor's hand near the tube, the motion of the falling plug can be stopped altogether.

Science Concepts

Electromagnetic induction was discovered in 1831 independently (and nearly simultaneously) by the American physicist Joseph Henry and the English physicist-chemist Michael Faraday. The principle that an electrical current and a corresponding magnetic field are created whenever there is relative motion between a conductor and a magnet is the basis for both electric generators and motors (which alternately either input mechanical energy and output electricity or the reverse). Interestingly, Henry and Faraday purposely did not file for patents on their inventions because they felt the ideas belonged to all humanity. In 1834, Russian physicist Heinrich Lenz discovered that the direction of the induced current (and the corresponding magnetic field) in a conductor always opposes the motion that produced it.

In this activity, as the magnetic slug moves down the copper tube, a series of transient "eddy currents" are induced in the copper tube, and these electrical currents create a magnetic field that opposes the downward fall of the magnetic slug. Some amusement park thrill rides use this same principle for braking, and some coin-operated vending machines use this principle for detecting counterfeit coins.

Science Education Concepts

This is a *visual participatory analogy* for the idea that if students are expected to truly understand (versus simply memorize and regurgitate) fundamental science concepts, teachers need to design curricular sequences that allocate *adequate time* for students to play with and think about phenomena so they can assimilate and accommodate the new ideas into their preexisting mental schema. This science "magic trick" can be used to develop scientific skills of inquiry and to emphasize to teachers that quickly *covering* content is unlikely to *uncover* prior misconceptions or help students construct a deep and lasting understanding of core ideas. "Less is more" and "learning takes time" are key recommendations of both the AAAS Benchmarks and the NRC National Science Education Standards (AAAS 1993; NRC 1996).

This activity also could be used to teach teachers the importance of "wait time" or the "pause-to-ponder principle" after they have asked questions that require higher-order thinking skills and depth of understanding. Alternatively, it can be used as a model to suggest that different students experiencing the "same" instructional activity will face different cognitive loads and require different conceptual processing times and instructional scaffolding.

Materials

- 1, 28–100 cm copper or aluminum tube (copper is preferred)
- 1 strong neodymium magnetic plug
- 1 brass plug (similar in appearance to the magnetic plug)
  (Both plugs should be just slightly smaller than the diameter of the tube. The tubes and magnetic and nonmagnetic plugs can be purchased independently from a variety of sources or as a packaged Eddy Current Kit. If assembling your own, use longer, heavier gauge and better conducting metal tubes and stronger magnets; this combination makes for the longest fall times.) Possible commercial sources include the following:
  - Arbor Scientific. *http://arborsci.co.* 800-367-6695. Lenz's Law Apparatus (has an uncovered copper tube with a cutaway viewing window). #P8-8400. $18. See also Arbor Scientific's Cool Stuff Newsletter Archive (*www.arborsci.com/CoolStuff/Archives.htm*) for lots of related teaching ideas.
  - Educational Innovations. *www.teachersource.com.* 888-912-7474. #M-200 for $12.95 or #M-225 (with slit) for $14.95. Both include a ~33 cm tube and two identical-looking slugs (for a 3–4 sec. fall through time. The copper tube in the kit without a slit has the explanation of the phenomenon printed on an outside label; you will want to cover the label if you intend to pass the tube around for examination or you can purchase the tube with the slit and no cover).
  - Penguin Magic. *www.penguinmagic.com.* 800-717-2799. Newton's Nightmare: $15.00: 10 in. blue aluminum tube with two plugs (magnet has ~5 sec. fall time) and four viewing holes.
- PVC, cardboard, or other nonconducting tubes of similar length and diameter can be used to demonstrate the importance of electrical conductivity to the effect.

Safety Note

Students and teacher should wear safety glasses or goggles during this activity.

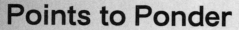

Points to Ponder

If a man begins with certainties, he shall end in doubts; but if he will be content to begin with doubts, he shall end in certainties…. To know truly is to know by causes.

—Francis Bacon, English essayist and philosopher (1561–1626)

When asked, What is the use of this discovery [electromagnetic induction], Faraday answered: *What is the use of a child—it grows to be a man.* Also from Faraday: *Nothing is too wonderful to be true if it be consistent with the laws of nature.*

—Michael Faraday, English physicist and chemist (1791–1867)

Procedure

This demonstration can be performed in the style of a sleight of hand magic trick (as in steps #1 and #2) or as a more straightforward study of systematic testing of several variables. It can be used as a general introduction to scientific inquiry or more specifically in a unit on electromagnetism and the design of motors and generators.

1. Introduce the demonstration with some verbal misdirection: "Today, we are going to be studying the universality and constancy of gravitational forces." Proceed to drop the nonmagnetic plug into the tube and catch it when it rapidly falls through (in a fraction of a second) as expected.

2. Repeat the experiment, stating, "Reproducibility is a key to good science," but secretly substitute the identical looking magnetic plug. Express feigned surprise with the greatly delayed result. If desired, repeat the "slow" and "fast" tests one or more times in random order to create a discrepancy. Challenge the learners to develop a list of probing questions and ideas they would like to test out to explore this phenomenon.

As with all magic tricks, there are a wide variety of "stories" that can be used to create a context that distracts the audience from paying careful attention or that deliberately misdirects the attention of the audience. In this case, one could be up-front about using two different metal plugs but indicate that the second (i.e., the neodymium magnet) is made from Moon rocks. Ask: How is gravity on the Moon different from that on Earth? Could this be a factor in the current experiment? If desired, use a tube with several regularly spaced viewing holes drilled along its length or a continuous slit for an inside view of the slow fall.

3. Share the Francis Bacon quote and elicit hypotheses as to what is going on in this demonstration. If someone does not suggest that you are using sleight of hand to switch between two different plugs, reveal your trick, but challenge participants to describe how the two plugs might be different and ask if any of these potential differences would logically make a difference in the rate of fall. If no one mentions magnetism, ask: Would it matter if one of the plugs were magnetic? Demonstrate that the plug that falls slowly is magnetic. Are magnets attracted to copper (or aluminum)? Demonstrate that a *stationary* magnet is not attracted to the copper tube. Ask: Is this second demonstration a fair representation of the "trick"? If not, what is missing? Would the motion of the magnetic plug be slowed if it were dropped through a plastic, cardboard, or other nonmetallic tube? Demonstrate this. What property do metals share that nonmetals do not have that might be a factor in this experiment?

4. Share Michael Faraday's quote and ask: If we combine a magnet, an electrical conductor (such as a copper wire or tube), and motion, what form or forms of energy are generated? As a hint, suggest that the answer has something to do with your action in turning the lights off and on in the room. Ask the learners to think about the simple act of turning a light switch on and off and about where the energy to power the lights comes from.

5. Ask: What are one or more ways of increasing the fall time of a magnetic slug? What variables in this demonstration could be changed to increase the discrepancy?

6. An additional "trick" can be your totally stopping the motion of a falling plug by holding the other plug hidden in your hand next to the tube (i.e., it does not matter which metal plug is on the inside or the outside). The force of magnetic attraction between the two plugs will extend through the tube and stop the motion of the inside plug.

Debriefing

When Working With Teachers

Discuss how science "magic tricks" can create a need-to-know motivation and help activate and diagnostically assess students' prior conceptions and skills of inquiry. Also discuss the Science Education Concepts, above. It is important to note that core scientific theories related to concepts such as atoms and molecules, cells, evolution, and plate tectonics took hundreds of years of development to reach their present forms. Expecting students to leap these monuments in a single bound is not realistic, nor does it do justice to the nature of science as a field of inquiry. Careful sequencing of concepts within and across units in individual courses and across the K–12 grades leads to deep understanding (in place of superficial memorization of factoids with a short half-life of memory). As a case in point, the fact that teachers themselves may not understand the basic principles of electric power generation should lead them to question the value of trying to cover too many concepts in too little time (and, conversely, to appreciate the wisdom of uncovering a smaller number of core concepts in greater depth). If desired, teachers can explore the research on student misconceptions about electricity and magnetism (e.g., see Driver et al. 1994; see chapters 15 and 16 as starting points). Teachers can also form study groups on the topics of cognitive load theory and questioning strategies (see Internet Connections for links to these topics).

When Working With Students

This magic trick can be followed by hands-on explorations in which students construct and study simple electric motors, generators, and electromagnets and use computer simulations that enable them to "see" invisible electric and magnetic fields.

Extensions

Students can work with handheld electric generators and/or motors; see Activity #18 in this book. Also see Internet Connections.

Internet Connections

- Arbor Scientific's Cool Stuff Newsletter: *www.arborsci.com/ CoolStuff/Archives3.aspx.* (See electricity, magnetism, and cool stuff #17 and #38 for lists of demonstrations.)

- Bloom's Taxonomy: *www.krummefamily.org/guides/bloom.html* (Revised and updated summary chart: *http://coe.sdsu/eet/articles/ bloomrev*)

- Changing Minds: Questioning Techniques (19): *http:// changingminds.org/techniques/questioning/questioning.htm*

- Exploratorium Snackbook: Eddy currents: *www.exploratorium. edu/snacks/eddy_currents*

- HyperPhysics: Faraday's and Lenz's laws (high school–college level explanations): *http://hyperphysics.phy-astr.gsu.edu/hbase/ electric/farlaw.html*

- Molecular Expressions: Interactive Java Lenz's law tutorial: *http://micro.magnet.fsu.edu/electromag/java/lenzlaw*

- Nondestructive Testing Resource Center's (NDT): Lenz's law applications: *www.ndt-ed.org/TeachingResources/NDT_Tips/LenzLaw. htm* (demonstrations) and *www.ndt-ed.org/AboutNDT/Selected Applications/Boeing-Liberty%20Bell/NDT-%20Liberty%20Bell.htm* (use of eddy currents to study the crack in the Liberty Bell)

- P16 Science Education at the Akron Global Polymer Academy: Wait time: *www.agpa.uakron.edu/p16/btp.php?id=wait-time*

- PBS Teachers Resource Roundups: Developing scientific thinking (pdf download): *www.pbs.org/teachers/resourceroundups*

- PhET Interactive Simulations: Faraday's law and electromagnetic lab: *http://phet.colorado.edu/simulations/sims. php?sim=Faradays_Law* and *http://phet.colorado.edu/simulations/ sims.php?sim=Faradays_Electromagnetic_Lab*

- University of Iowa Physics and Astronomy Demonstrations (video clips): Electricity and magnetism: *http://faraday.physics. uiowa.edu* (seven demonstrations on eddy currents).

- University of Victoria: Bloom's Taxonomy: *www.coun.uvic.ca/ learning/exams/blooms-taxonomy.html*

- University of Virginia Physics Department: Experiments with electromagnets: *http://galileo.phys.virginia.edu/ outreach/8thGradeSOL/ElectromagnetFrm.htm*

- Wikipedia: (1) Bloom: *http://en.wikipedia.org/wiki/Taxonomy_ of_Educational_Objectives* (2) Cognitive load theory: *http:// en.wikipedia.org/wiki/Cognitive_load_theory* (3) Eddy currents: *http://en.wikipedia.org/wiki/Eddy_current* (4) Lenz's law: *http:// en.wikipedia.org/wiki/Lenz%27s_law* (5) Michael Faraday: *http:// en.wikipedia.org/wiki/Michael_Faraday*

Answers to Questions in Procedure, steps #2–#5

2. The gravity on the surface of the Moon is approximately one-sixth that on Earth. However, even if the metal slug were made from a Moon rock (which it is not), gravity is always an interaction between two masses, not an intrinsic property of a single mass. A Moon rock would fall under the Earth's gravitational attraction at the same rate (9.8 m/s²) as any object that was from Earth (discounting any possible air resistance effect).

3. Possible differences between the two plugs could be that the plugs are of different masses (but all masses fall at the same rate unless air resistance is a significant factor) or different sizes (show the class that this is not the case and, even if it were, it would only be a factor for a light object that experienced significantly different amounts of air resistance). Also, one plug might be magnetic (but *stationary* magnets are not attracted to copper—or aluminum). Of course, this demonstration involves a moving magnet falling through an electrical conductor and that combination is the key to the solving the mystery or explaining the magic. The motion of the magnetic plug will not be slowed when it is dropped through a plastic, cardboard, or other nonconducting tube of the same diameter and length as the metal tube.

4. When one combines a magnet, an electrical conductor, and motion, kinetic energy is converted into electricity and magnetic fields—this is the principle of the electric dynamo or generator that provides nearly all the electricity produced worldwide (electrochemical cells/batteries, fuel cells, and photovoltaic cells produce only a minuscule amount of the world's total electric power). In electric power plants, the source of the kinetic energy can come from water that is heated into high pressure steam with fossil fuels, nuclear power, or concentrated solar energy; the kinetic energy also can come directly from falling water (hydroelectric) or blowing wind (windmills).

5. The time it takes the plug to fall through the tube can be increased by using a stronger magnet (e.g., neodymium), a longer tube, and/or a tube made of a better electrical conducting metal (e.g., copper versus aluminum versus iron).

Cognitive Inertia:
Seeking Conceptual Change

Expected Outcome

Two identical strings are attached to a kilogram mass. The first string is given a quick jerk and breaks; the second is pulled slowly and is able to move and lift the mass. Also, a pen balanced on a knitting hoop resting on a narrow-mouth bottle is observed to drop straight down into the bottle when the hoop is hit horizontally from inside the circular hoop.

Science Concepts

Inertia, a principle that was explored by Galileo and codified in Newton's first law, is the tendency of a body at rest to remain at rest (and the tendency of a body in motion to continue in straight line, nonaccelerated motion). The second activity with the hoop also involves the independence of horizontal and vertical forces.

Science Education Concepts

Students' preinstructional conceptions may be erroneous and quite resistant to change. Cognitive inertia (or conservatism)—the tendency of humans to continue to think both what and how they have previously thought—applies at both the individual and the scientific community level. From a teaching perspective, this resistance to change (or cognitive conservatism) can be overcome if discrepant events are used to activate and challenge misconceptions in planned instructional sequences (e.g., the 5E Teaching Cycle; see Appendix B for more information on this teaching cycle). Such sequences create optimal cognitive loads and learning challenges by giving students adequate time to construct new, improved, and expanded conceptual networks of understanding based on empirical data, logical argument, and skeptical review.

Materials

Experiment #1
- A kilogram mass (or a heavy science textbook used by the class!)
- 2, 14 in. pieces of string strong enough to individually support the mass, but able to be snapped if pulled suddenly. Some trial and error may be needed to find an optimal string.

Experiment #2
- 1, 10–14 in. diameter circular knitting hoop
- 1 pen with a flat bottom (so it can stand on end)
- 1 glass bottle with a mouth not much wider than the diameter of the pen (so that the pen can fall through the mouth, such as a 16 oz. soft drink bottle)
- 1 glass bottle with a much wider mouth than the one listed above (2 in. or greater, such as a gallon jug)

Safety Notes

1. Students and teacher should wear safety glasses or goggles during this activity.
2. Use proper foot protection (closed-toe shoes or sneakers only).

Points to Ponder

For it is too bad that there are so few who seek the truth and so few who do not follow a mistaken method in philosophy. This is not, however, the place to lament the misery of our century, but to rejoice with you over such beautiful ideas for proving the truth.

—Galileo Galilei, Italian astronomer-physicist (1564–1642), in a letter to Johannes Kepler, German astronomer (1571–1630)

If I have seen a little further it is by standing on the shoulders of Giants…. I do not know what I may appear to the world…. To myself I seem to have been only like a boy playing on the seashore, and diverting myself in now and then finding a smoother pebble or a prettier shell than ordinary, whilst the great ocean of truth lay all undiscovered before me.

—Isaac Newton, English physicist-mathematician (1642–1727)

Procedure

(See answers to the questions in Experiments #1 and #2 on p. 258.)

Experiment #1: *Slow and Steady Wins the Day*

1. *Ask:* Do you think that I can move this 1 kg mass (or heavy science textbook) by pulling on one of these two identical strings? What variables might determine my success or failure? *Demonstrate:* Give one of the strings a quick jerk and the string will break without moving the mass. Why did the mass resist being moved? Is the resistance to motion only due to friction with the tabletop? Test this by suspending the mass in a vertical direction with one string and give the second string a quick jerk in the horizontal direction.

2. *Ask:* Would an object on a space shuttle experiencing zero gravity resist being moved? Would it be easier or harder if the object has a mass of 0.5 kg? 2.0 kg? How can I simulate zero-gravity in the classroom? Is there a way I can succeed in moving the mass using the second string? *Demonstrate:* Give the second, remaining

string a slow, steady pull and you can move and even suspend the 1 kg mass in air.

3. *A "magic" variation on this demonstration:* A round "inertia ball" or cylindrical metal weight with two rings or hooks at 180 degrees from each other can be suspended vertically by one string with the second string hanging freely down from the other ring. The top or the bottom string will break depending on whether you give the bottom string a slow steady pull or a quick jerk. You can use the misdirection ploy that magicians use and suggest that your verbal command is the key.

Experiment #2: *"Hooping" the Pen Will Fall Into the Bottle*

Demonstrate that a pen (with cap removed) will fit into the mouth of the first bottle without a lot of room to spare but easily fits into the second, wider mouth bottle. Ask a volunteer to attempt to drop the pen into the narrow-mouth bottle from a height of 1 ft. The volunteer will probably not meet with success. Suggest that as a science teacher, you can do this in an even more interesting way. Balance a knitting hoop on top of the first, narrow-mouth bottle and carefully balance the vertically placed pen on top of the hoop, centered just over the mouth of the bottle (this will take a little practice and steady hands). Place your right hand into the middle of the hoop with your palm facing to the left side of the hoop (or reverse directions if you want to use your left hand). Then, quickly slap your right hand horizontally to the left. With a little prior practice you can consistently avoid giving the hoop any vertical motion and the hoop will move out from under the pen, leaving it seemingly suspended in mid-air before it falls straight down into the glass bottle (somewhat akin to Wile E. Coyote in the Road Runner cartoons). (*Note:* The visual effect of seemingly suspended animation is accented by the fact that falling objects accelerate ($9.8 \ m/s^2$) as they fall, which means that they start off motionless and then, in equal time intervals, they fall progressively farther distances.)

Ask for a student volunteer to repeat the trick. Ask if he or she would like to use the wider-mouth bottle for practice. Ask the class: What are the advantages of using a wider-mouth bottle? Are there any disadvantages of a using a wider bottle? If so, what are they and why are they disadvantages? If desired, you can take this activity to the extreme by using a jar with a 5–6 in. opening, which will allow most students to easily see the problem with wide-mouth containers.

Debriefing

When Working With Teachers

The two demonstrations (Experiments #1 and #2) can be used as visual participatory analogies for the idea of cognitive load and inertia and the need for the teacher to provide adequate time and instructional scaffolding to help students achieve depth of understanding.

The first demonstration suggests that teachers need to take time (i.e., a slow steady pull versus a quick jerk) to help students revise (or replace) prior conceptions that run counter to accepted science and to acquire new, scientifically more accurate conceptions. Most big ideas in science developed over hundreds of years, yet we expect students to rapidly understand the ideas, without the benefit of empirical evidence, logical argument, and skeptical review for how we know what we know. This is why the AAAS's Benchmarks and the NRC's National Science Education Standards (AAA 1993; NRC 1996) call for a less-is-more approach with respect to content coverage.

Similarly, students need to make regular, steady efforts to learn the science in their textbooks rather than using the quick-jerk, cramming-before-the-test approach. In the education literature this is referred to as massed practice versus spaced or distributed practice (see Internet Connections: Wikipedia: Spacing Effect) and can be related to how athletic teams prepare for a big games (i.e., they never pull all-nighters before a big game).

The second demonstration points out that what may seem like a difficult task (i.e., learning science) can be accomplished if you exert your efforts in the right direction and invest in sustained practice over time for consistency in performance. The demonstration also serves as a visual analogy for instructional approaches that pull out the scaffolded support from students too quickly and have them "free fall" into a test. Teachers can explore the Wikipedia and the NARST websites (see Internet Connections) for the related educational theories and research on student misconceptions regarding forces, horizontal motion, and gravity (see Driver et al. 1994, chapters 21–23, for an overview).

When Working With Students

These Engage-phase activities can be used near the start of a unit dealing with Newton's laws of motion. Initially, both activities can be presented in a magic style, with the explanations withheld until after students have experienced hands-on Explore-phase activities. Then, when the Explain phase is introduced, students should be more ready to figure out the trick for themselves. When revisiting the phenomenon, challenge students to predict-observe-explain the different outcomes that result from hitting the hoop horizontally from the outside in versus from the inside out. Also ask students to draw contrasting images of how the hoop distorts when you hit it from the outside (horizontal compression and vertical elongation, which cause the hoop to hit the mouth of the bottle) versus the inside (horizontal elongation and vertical compression, which reduce contact and friction with the mouth of the bottle and allow the force to be directed horizontally without a significant vertical component).

Extensions

Science demonstration books and internet sites contain a wide variety of hands-on Exploration-phase activities and classic demonstrations on the concept of inertia (see Internet Connections).

Internet Connections

- American Educator. Ask a Cognitive Scientist column: *www. danielwillingham.com* (1) Practice makes perfect—But only if you practice beyond the point of perfection: *www.aft.org/pubs-reports/american_educator/spring2004/cogsci.html* (2) Why students think they understand—When they don't: *www.aft.org/pubs-reports/american_educator/winter03-04/cognitive.html*

- Arbor Scientific's Cool Stuff Newsletter: *www.arborsci.com/CoolStuff/Archives3.aspx.* (See Forces and Motion listing for inertia demonstrations.)

- National Association for Research in Science Teaching (NARST): Research matters to the science teacher: Over 30 synopses including authentic science teaching, constructivism, higher order cognitive skills, conceptual change teaching, and using questions. *http://narst.org/publications/research.cfm*

- North Carolina Science Teachers Association: The Science Reflector: 10 Inertia experiments and demonstrations: *http://faculty.uncfsu.edu/grahi/inertiaexp.pdf*

- University of Iowa Physics and Astronomy Lecture Demonstrations (video clips): Mechanics: *http://faraday.physics.uiowa.edu* (seven demonstrations on inertia of rest) and mass and string "magic" demonstration: *http://faraday.physics.uiowa.edu/mech/1F20.10.htm*

- University of Virginia Phun Physics Show (three inertia demonstrations including the hoop and pen): *http://phun.physics.virginia.edu/topics/inertia.html*

- University of Virginia Physics Department: Newton's first law: Observing inertia (demos and HOEs): *http://galileo.phys.virginia.edu/outreach/8thGradeSOL/Newton1Frm.htm*

- University of Wisconsin-Madison Physics Lecture Demonstrations (large sample of Newton's first law): *http://demo1.physics.wisc.edu/UW-Demos_Pira/1F-NewtonsFirstLaw.html*

- Wake Forest University: Newton's laws videos: Tablecloth pull and toilet paper: *www.wfu.edu/physics/demolabs/demos/avimov/bychptr/chptr2_newton.htm*

- Wikipedia: (1) Cognitive load theory: *http://en.wikipedia.org/wiki/Cognitive_load_theory* (2) List of cognitive biases: *http://en.wikipedia.org/wiki/Cognitive_biases* (3) Spacing effect: *http://en.wikipedia.org/wiki/Spacing_effect* (4) Zone of proximal development: *http://en.wikipedia.org/wiki/Zone_of_Proximal_Development*

Answers to Questions in Procedure, Experiment #1 and Experiment #2

Experiment #1: Any stationary object has a resistance to motion by virtue of its inertial mass. Additionally, objects at rest on a surface or suspended in a fluid (liquid or gas) will encounter frictional forces that make them harder to move. When the string is pulled quickly, both of these factors result in the string breaking. However, a slower, steady pull will enable the mass to overcome these factors and move without putting too much tension on the string. In fact, the mass can be lifted off the table and suspended in air without the string breaking. Objects in the vacuum of space do not experience frictional forces since they are not in a fluid or resting on a solid base, but they still have inertial mass and resist a change in motion. A free-fall demonstration is a way to simulate microgravity in a classroom.

Experiment #2: Most people will think that a wider-mouth bottle makes the trick easier to perform, but it actually makes it nearly impossible. If the hoop curves downward into the wider mouth, any horizontal motion or slap will develop a vertical component that will cause the pen to be jerked in a tumbling motion and put it off-center for dropping into the bottle.

Activity 25

Optics and Mirrors: Challenging Learners' Illusions

Expected Outcome

Coins dropped into a magic bank appear to shrink in size and fall through a narrow funnel into a tiny chamber or disappear altogether (depending on which type of magic bank is used).

Science Concepts

Science depends on empirical evidence, logical argument, and skeptical review. Optical illusions challenge us to consider if our eyes sometimes play tricks on us. Furthermore, learning requires mental activity on the part of the learner; understanding is not something that can be passively absorbed or done for the learner by the teacher. (Teachers at the high school level can have students take apart the magic banks and/or construct models that replicate the bank's discrepant behavior to discover the underlying principles of optics and mirrors.)

Science Education Concepts

Beyond providing an insight into the nature of science, this activity serves as a *visual participatory analogy* to challenge the validity of the "banking" model of teaching and learning. The misconception that the quality of learning is largely a function of what the teacher deposits into the empty vessel of the student's mind is held by students of all ages. The truth is quite different: *Learners* (a more active term than *students*) will actively invest energy and effort into instructional activities in proportion to the expected return on their investments. The expectancy-value theory of motivation suggests that the perceived "pain" or "costs" of learning efforts must be counterbalanced by the expected "gains." In addition, the psychological rewards of learning science should be even greater than the performance rewards (earning acceptable or even good grades). Effective science instruction empowers learners to construct personally meaningful understandings about and appreciation for the nature of science.

Materials

Toy and novelty stores sell a variety of types of magic banks that rely on optical illusions to work their magic. Magic banks have even been given away as toys in fast food restaurant kids' meals. Here is one source of magic banks:

- Educational Innovations. *www.teachersource.com.* 888-912-7474. Sells two varieties that both hold up to 50 coins: Shrinking Coin Bank (3.1 in. × 3.1 in. × 4.8 in.). BNK-100. $8.95, and Disappearing Coin Bank (2.75 in. cube). #BNK-205. $5.50.

Points to Ponder

There are two modes of acquiring knowledge, namely by reasoning and experience. Reasoning draws a conclusion and makes us grant the conclusion, but does not make the conclusion certain, nor does it remove doubt so that the mind may rest on the intuition of truth, unless the mind discovers it by the path of experience.

—Roger Bacon, English natural philosopher-clergyman (1220–1292)

The mind of man is being habituated by a new method and ideal: There is but one sure road of access to truth—the road of experiment, record and controlled reflection.

—John Dewey, American philosopher-educator (1859–1952)

Procedure

This activity can be used as an introduction to a unit on the nature of science or on optics or it can be used to emphasize the active nature of learning. (See answers to the questions in steps #1–#4 on pp. 264–265.)

1. Circulate several magic banks and a bag of pennies and ask the learners to deposit one penny in a bank and observe what happens. (The pennies dropped into the magic banks appear to shrink in size and fall through a narrow funnel into a tiny chamber or disappear altogether, depending on which type of magic bank is used.)

    Ask: From a science perspective, is seeing always believing? How do different scientific instruments, procedures (such as dissection), and theories probe beneath the surface of phenomena to get at underlying explanations? In this particular case, what concepts would need to be explored in order to understand the science behind the magic trick? What indirect evidence could we use to support the idea that the penny has not really disappeared?

2. Ask: Is learning something that a teacher can do for (or to) a student? Is it possible for knowledge and understanding to be transmitted by a teacher to a learner who needs only to be tuned in to the right frequency, pay attention, and passively receive the broadcasted signal? Who can make "deposits" into the "bank" of a learner's mind?

3. Compare the quotes of the Roger Bacon and John Dewey, two scholars separated by nearly 600 years. Would either of them accept the "banking model" of learning? Would they want the teacher to give students the answer to the puzzle of the magic bank or would they think students should (at least partially) discover the answer on their own through experimentation?

4. Consider another analogy between banking and learning: How can students' efforts to learn be considered an investment? What motivates students to take the risk of investing their time and energy in learning science? What potential rewards (psychological and other) can students expect as a likely return on their investments? Are grades a form of dividends or payments for learning, or should the reward consist of enhanced understanding of interesting, relevant science phenomena (or perhaps a mix of the two)? Do most students feel that the gain or benefit of learning science is greater than the pain or cost of the work they need to do? If not, why?

Debriefing

When Working With Teachers

Emphasize the analogy of the magic bank to theories about learning as discussed in the Science Education Concepts section. The Internet Connections section provides links to background readings on the cognitive science and constructivist theories of learning, including articles on the use of extrinsic rewards and praise to motivate effort and learning. Teachers can also explore the seminal writings of John Dewey. His books *How We Think: Democracy and Education* (1910/1933), *Experience and Nature* (1925), and *Experience and Education* (1938) all provide insights for modern day science teachers who want to employ hands-on, minds-on, and real-world approaches in their classrooms.

When Working With Students

Briefly discuss the students' perspectives on the questions in the Procedure section. If this activity is used during an introductory unit on the nature of science, consider having students explore optical illusions that can be resolved by making simple linear measurements (e.g., illusions related to the apparent sizes of lines or figures). If used during the Engage phase of a unit on optics, do not attempt to prematurely explain the optics behind the illusion, but rather use additional Engage- and Explore-phase activities to allow students to refine their questions before beginning Explain-phase activities (see Appendix B for a discussion of the 5E Teaching Cycle). Challenge students to build optical illusion boxes similar to the magic boxes, using plastic or mylar mirrors. Be sure to refer back to the visual metaphor of banking versus building throughout the course to empower learners to take charge of their own learning.

Extensions

The internet has a large number of sites with interactive illusions of various kinds. Some websites may contain pop-up ads and other material that may be inappropriate for students of certain ages, so be sure to review the sites before using them "live" in the classroom. Also see the websites listed in Activity #5 in this book.

Internet Connections

- American Educator, Ask a Cognitive Scientist column: *www.danielwillingham.com*: (1) How praise can motivate–Or stifle: *www.aft.org/pubs-reports/american_educator/issues/winter05-06/cogsci.htm* (2) Should learning be its own reward?: *www.aft.org/pubs-reports/american_educator/issues/winter07_08/scientist.htm* (3) Caution: Praise can be dangerous (download pdf article by Carol S. Dweck): *www.aft.org/pubs-reports/american_educator/spring99/index.html*

- Illusion Works (visual and some auditory illusions with explanations): *http://psylux.psych.tu-dresden.de/i1/kaw/diverses%20Material/www.illusionworks.com*

• Wikipedia: (1) Constructivism: *http://en.wikipedia.org/wiki/Constructivism_%28learning_theory%29* (2) Constructivist teaching: *http://en.wikipedia.org/wiki/Constructivist_teaching_methods* (3) Expectancy theory: *http://en.wikipedia.org/wiki/Expectancy_theory* (4) Expectancy-value theory: *http://en.wikipedia.org/wiki/Expectancy-value_theory* (5) John Dewey: *http://en.wikipedia.org/wiki/John_Dewey* (6) Roger Bacon: *http://en.wikipedia.org/wiki/Roger_Bacon*

Answers to Questions in Procedure, steps #1–#4

1. The human visual system *generally* works quickly, reliably, and (seemingly) effortlessly to give us operational descriptions of reality that enable us to adapt and survive to changing external realities (both as individuals and as a species). However, the inferences that the visual system makes automatically are subject to "miss-takes" and it can be tricked. Most students realize that magic is often about "smoke and mirrors"—or, in this case, about the clever use of optics and mirrors. Massing (or weighing) a penny and the bank before and after the trick would indirectly confirm that although the pennies are out of sight (or seem to have become smaller), they are still in the bank and remain the same size.

2. These questions are designed to challenge the notion that learning is something a teacher can do to or for a passive, receptive student and to help students see that learning is instead a process of active construction on the part of the student with scaffolded support from the teacher.

3. Bacon and Dewey were both advocates for scientific experimentation as means to advance individual and societal knowledge. Dewey was an early and strong advocate for science in the K–12 curriculum because of its ability to develop scientific habits of mind and problem-solving skills and for its role in advancing human well-being.

4. These questions are intended as open-ended discussion-starters. Cognitive scientists have discovered that intellectual challenges that are neither too hard nor too easy motivate effort and, when solved, elicit neurochemicals in the learner that produce feelings of pleasure and accomplishment, which in turn motivate the learner to tackle future challenges. This feedback loop of naturally produced "highs" gives new meaning to the phrase "your brain on drugs"! Perhaps part of reason for the illegal-drug problem in our schools is that students too infrequently receive the healthy, 100% natural biochemical boost that comes from successfully learning important concepts; solving interesting, minds-on problems; and completing challenging hands-on tasks.

Activity 26

Polarizing Filters:
Examining Our Conceptual Filters

Expected Outcome

Two polarizing filters (or two lenses from Polaroid sunglasses) are overlaid on an overhead projector. As the top filter is rotated, the amount of transmitted light varies from nearly as much as that allowed by one filter (or lens) to almost none (i.e., the filters appear opaque or black).

Science Concepts

Regular visible light is not polarized; the electromagnetic (EM) waves vibrate in all possible directions in 3D space. Light reflected off flat, horizontal, reflective surfaces (such as car hoods) becomes polarized in the horizontal direction and produces visual glare. Polaroid sunglasses reduce this glare by significantly blocking EM vibrations in the horizontal direction in a manner analogous to trying to vibrate a string horizontally through vertical slats. If two polarizing filters are placed perpendicular to each other, both horizontal and vertical vibrations will be blocked, allowing little light to be transmitted.

Science Education Concepts

This demonstration is a *visual participatory analogy* for how students come to "see" science concepts differently as their conceptual networks are modified as a result of new learning experiences. Misconceptions are rarely replaced instantaneously or completely in a single "aha!" moment of insight. When revising or replacing erroneous conceptions (or ineffective teaching practices), learners may experience an "implementation dip"—that is, they try to use some combination of old and new "lenses" (or conceptual filters) with variable success. Conceptual and behavioral changes take time, guided practice, and feedback; the changes depend on an often unconscious assessment that the psychological reward, benefit, or gain is worth the risk, pain, or cost of the change. "Miss-takes" are often necessary prerequisites for learning something that is challenging or for overcoming cognitive biases.

Materials

- 2 polarizing film/filters or lenses (from a pair of Polaroid sunglasses)
- 1 cheap (non-Polaroid) pair of sunglasses
- Overhead projector
- For a macroscopic model of polarization: vertical slats or a small section of a gate and a rope

- Inexpensive (plastic) linear polarizing film can be purchased from most science suppliers (in a variety of sizes):
  - Arbor Scientific. *http://arborsci.com.* 800-367-6695. #P2-9405. 50, 35 mm slide mounts, $33
  - Educational Innovations. *www.teachersource.com.* 888-912-7474. 1, 3 in. × 3 in. square, #PF-3 for $3.75 or a 6 in. × 6 in. square, #PF-4 for $10.50

Points to Ponder

It is one thing to show a man he is in error, and another to put him in the possession of the truth.

—John Locke, English philosopher (1632–1704) in *An Essay Concerning Human Understanding*

It is much easier to recognize error than to find truth; error is superficial and may be corrected; truth lies hidden in the depth…. To find a new truth there is nothing more harmful than an old error.

—Johann Wolfgang von Goethe, German poet-dramatist-novelist (1749–1832)

It's not what we don't know that hurts; it's what we know that ain't so.

—Will Rogers, American humorist (1879–1935)

Teach a highly educated person that it is not a disgrace to fail and that he must analyze every failure to find its cause. He must learn how to fail intelligently, for failing is one of the greatest arts in the world.

—Charles F. Kettering, American engineer-inventor (1876–1958)

Procedure

(Answers to questions in steps #1 and #2 are found on p. 273.)

1. Hold up an inexpensive (non-Polaroid) pair of sunglasses and ask: How do these sunglasses work? Place the sunglasses on the overhead projector to demonstrate how they work (in step #2).

2. What makes a more expensive pair of sunglasses, such as Polaroids, better and worth the extra expense? What benefits do advertisers claim for the more expensive glasses? Compare the light transmitted through two inexpensive lenses versus that transmitted through two Polaroid lenses (or polarizing filters), especially noting the effect when rotating the top lens over the bottom lens. (*Note:* You'll need two pairs of sunglasses to do this comparison unless you remove the lenses from one pair.)

3. If desired, a macroscopic model of polarization can be made by showing that a vertically oriented S-shaped standing wave made with a rope can be maintained through a section of a vertically oriented gate whereas a horizontally polarized, sideways S-shaped rope wave cannot. Also, if one gate is held perpendicular to a second, neither type of rope wave can pass; this is analogous to using two Polaroid lenses (or polarizing filters) at a 90 degree overlap.

Debriefing

When Working With Teachers

Use the demonstration as a visual analogy for the preexisting conceptual filters or lenses we all wear that are based on our individual processing of our previous life experiences (both formal schooling and general life circumstances). Our prior conceptions influence how we perceive and process new experiences. Misconceptions may cause us to make certain consistent errors or "miss-takes" unless they are corrected. Students (and teachers and scientists) will "switch glasses" and give up their "old glasses" (i.e., previous ways of seeing and behaving) only if they have seen that the old glasses do not work as well as the new, improved prescription glasses, which typically take a little getting used to.

Well selected and sequenced discrepant-event activities are effective in motivating and helping students (and teachers) to make conceptual and behavioral changes, but this process takes time. In another analogy to learning, contrast the short- and long-term efficacy of painting a rusting metal door if you take the time to remove the rust first versus if you do not. (*Note:* One weakness of this analogy is that misconceptions are not removed all at once; people may merely restrict the range of ideas to which they apply faulty concepts). Link this discussion to one or more of the quotes.

When Working With Students

This demonstration can be used (as with teachers; see p. 270) to discuss the challenge of conceptual change in learning science (and the challenge of unlearning prior misconceptions). As a science demonstration, the activity would fall in a unit on light, optics, and lenses. The macroscopic model makes a nice Explain-phase activity (see Appendix B for a discussion of the 5E Teaching Cycle).

Extensions

1. *Eye and Skin Protection Experiments.* Ultraviolet radiation presents a critical health and safety issue (especially given the human-induced reduction of the Earth's ozone layer in the stratosphere). Inexpensive sunglasses reduce visible light but typically do not block harmful UV (ultraviolet) rays. In fact, the iris of the eye responds to reduced light by opening slightly wider, allowing more of the harmful UV rays to reach the eye than if no sunglasses were used at all! By contrast, blue blocker sunglasses filter out much of the higher frequency, higher energy blue, violet, and ultraviolet light. This filtering both protects the eyes and accents orange and yellow colors. If a blue transparency is placed on an overhead projector (or projected from a PowerPoint slide), it will appear almost black when viewed through blue blocker sunglasses. Both of the science suppliers that appear in the Materials list sell a variety of UV-sensitive, color-changing beads that can be used to test the relative UV-blocking ability of different sunglasses (and the relative effectiveness of different SPF-rated sunscreens). These beads can be used for a variety of student-designed experiments.

2. *Polarizing Patterns Predict Properties.* Various optically active materials (e.g., rigid, translucent plastics, plastic utensils, cellophane tape, polyethylene, corn syrup, honey, mica, and calcite crystals) reveal unique color patterns when placed between two rotating Polaroid filters. Scientists use this technique to reveal stress points, measure thickness, and probe into molecular structures. Science suppliers such as Educational Innovations sell kits to demonstrate these patterns or you can assemble the materials yourself. (See Internet Connections: Exploratorium Science: Polarized Light Mosaic.)

Internet Connections

- Arbor Scientific's Cool Stuff Newsletter: *www.arborsci.com/Cool Stuff/Archives3.aspx* (See Optics: Polarization in art and other polarization demonstrations.)

- Austine Studios Polarized Light Art: *www.austine.com*

- Brigham Young University Physics Computer Resources: select Optics: Polarize: *http://stokes.byu.edu/computer_resources.html*

- Exploratorium Science Snackbook: Polarized light mosaic: *www.exploratorium.edu/snacks/polarized_mosaic* and Polarized sunglasses: *www.exploratorium.edu/snacks/polarized_sunglasses*

- HowStuffWorks: Sunglasses: *http://science.howstuffworks.com/sunglass.htm*

- University of Iowa Physics and Astronomy Lecture Demonstrations (video clips): Optics: *http://faraday.physics.uiowa.edu* (13 demonstrations on polarization)

- University of Virginia Physics Department: HOEs with polarized filters and light: *http://galileo.phys.virginia.edu/outreach/8thGradeSOL/PolarizedFiltersFrm.htm* and *http://galileo.phys.virginia.edu/outreach/8thGradeSOL/PolarizedLightFrm.htm*

- Wikipedia: (1) List of cognitive biases: *http://en.wikipedia.org/wiki/Cognitive_biases* (2) Ozone layer: http://en.wikipedia.org/wiki/Ozone_layer (3) Polarization: *http://en.wikipedia.org/wiki/Polarization* (4) Sunglasses: *http://en.wikipedia.org/wiki/Sunglasses* (5) Sunscreen and SPF: *http://en.wikipedia.org/wiki/Sunscreen#Sun_protection_factor*

Answers to Questions in Procedure, steps #1 and #2

1. Non-Polaroid, color-tinted sunglasses absorb and thereby filter out some sunlight before it reaches the eye.

2. Polaroid lenses reduce road glare by greatly reducing horizontally polarized light reflected off concrete roads, automobiles, and snow on flat ground. Light coming through two inexpensive lenses will not vary when the top lens is rotated; the two overlapping Polaroid lenses, on the other hand, will appear much darker in one position. Polaroid sunglasses are designed to reduce horizontal reflective glare while allowing vertically polarized light to pass through to the eye.

Activity 27

Invisible Gases Matter: Knowledge Pours Poorly

2 HOLE STOPPER (BOTH OPEN)

2 HOLE STOPPER (1 PLUGGED)

Expected Outcome

Water flows down through two identical funnels, each inserted in a two-hole stopper, into two identical flasks. The water flows at very different rates, however, because a clay plug has been hidden in one of the holes in one two-hole stopper. In the slow, oddly behaving flask, some of the water actually shoots back out!

Science Concept

Invisible gases are a form of matter that have volume or occupy space.

Science Education Concepts

The notion that students are empty receptacles ready to receive knowledge poured into their heads by teachers persists despite words to the contrary passed down from ancient philosopher-teachers and despite decades of modern-day cognitive science research that points to the fallacy of the "empty vessel."

This demonstration serves as a *visual participatory analogy* to challenge the "unquestioned answers" that teachers have traditionally learned themselves and passed on to their students. The demonstration also highlights the importance of teaching that catalyzes minds-on learning by activating students' attention and challenging their prior knowledge (i.e., in many cases, what the student already "knows" may not be so). Engaging learners in "dialogue" with natural phenomena, one another, their own prior conceptions, and the teacher is a core principle of constructivist teaching.

Materials

- 2 identical size (500 mL) Erlenmeyer flasks (or bottles)
- 2 glass funnels
- 2 two-hole rubber stoppers
- 2 clear, colorless plastic cups filled with water
- 2 colors of food coloring
- Small piece of clay

Safety Notes

1. Students and teacher should wear indirectly vented chemical splash goggles during this activity.

2. Use caution when inserting glass funnel into stopper (Procedure, step #1). The funnel can break, creating a sharp that can puncture the skin.

Prior to the activity, use the clay to plug one of the holes on one of the two-hole stoppers. Put the clay in the middle of the hole so that the plug is not readily observed without looking directly into the hole. *This is one of the few activities in this book in which the teacher uses deliberate deception to (temporarily) "trick" students or distract their attention from a key observation.* Alternatively, you can use a one-hole stopper if you want a more readily noticeable difference (the funnel will be clearly centered in a one-hole stopper versus in a two-hole stopper, where it will be off center). A white backdrop and two boxes for elevation heighten the effect of this demonstration.

276

Points to Ponder

The most useful piece of learning for the uses of life is to unlearn what is untrue.

—Antisthenes, Greek comic dramatist (445–365 BC)

What we have to learn to do, we learn by doing.

—Aristotle, Greek philosopher (384–322 BC)

Iron rusts from disuse, stagnant water loses its purity, and in cold weather becomes frozen; even so does inaction sap the vigors of the mind.

—Leonardo da Vinci, Italian polymath (1452–1519)

Young people are not receptacles to be filled; they are fires to be kindled.

—Justus von Liebig, German chemist (1803–1873)

Procedure

1. Before learners arrive, insert a glass funnel into each of the two-hole stoppers and firmly insert the stoppers into the Erlenmeyer flasks to prepare what appear to be two identical systems.

2a. *When Working With Teachers*: Introduce the demonstration by suggesting that based on years of being students, we all have acquired many tacit and unquestioned assumptions about teaching and learning science and that in this activity we will be using colored liquids to represent knowledge and an Erlenmeyer flask–funnel system to represent the learner.

2b. *When Working With Students:* Introduce the demonstration as a means to begin to explore the properties of various states of matter (i.e., gases are not "no thing" but something with tangible properties).

3. Ask for a volunteer to assist you in front of the group. Ask the volunteer to pick his or her favorite food coloring and to dye his or her cup of water; as the teacher, you do the same thing with

the second cup. Explain that the challenge in this demonstration is to see who can most quickly transfer his or her cup of water (knowledge) into the Erlenmeyer flask (student mind) through the funnel. (*For Teachers Only:* Note the visual appeal of using colored water [i.e., the very act of dropping colored dye into the water attracts attention and may even raise questions about the nature of diffusion and mixtures]. Given international competition in a global "flat world" economy and the accelerating pace of sci-tech discoveries and applications, education can be viewed analogically as a "race" where losing nations are "at risk.")

4. Without drawing attention to the fact that two-hole stoppers are being used, place the two setups on two elevated boxes in front of a white backdrop and ask the volunteer assistant to stand on one side. You stand on the other. Both of you begin pouring water at the same time. It is advantageous (though not essential) that the teacher select the rigged setup with the clay plug.

5. (See answers to questions in this step on pp. 282–283.) Ask: How might we account for the difference in behaviors of the two setups? Do you think this was a fair test? What allows the water to easily drop into the volunteer's setup? How might we explain the unusual behavior of the instructor's setup? Is either of the flasks really "empty" at the start of the demonstration? If not, what are they filled with?

6. Ask the volunteer to take a closer look at the two setups and describe what he or she sees; the plug in one of the two holes in the instructor's setup should be noticed. The teacher's setup will work more like the volunteer's setup if a straw with a smaller diameter than the funnel's stem is inserted through the funnel with the straw's top end initially covered with the index finger and then released once it is through the funnel. This essentially creates a two-hole system.

Debriefing

When Working With Teachers

Critique the "fit" of the visual analogy. (See answers to questions #1–#4 on p. 283.)

1. Are learners' minds of fixed, limited capacity in terms of their potential to learn?

2. Are learners' minds empty at the start of the race?

3. Are learners' minds chiseled in stone? (Does new information get passively added on or does it interact with prior knowledge?)

4. Are learners' minds receptacles to be filled or fires to be kindled?

Teachers can be encouraged to explore how cognitive science and the learning theory of constructivism apply to science teaching (see Internet Connections). Teachers will also want to delve into the following valuable references on the nature of student misconceptions and their implications for teaching: Driver, Guesne, and Tiberghein 1985; Driver et al. 1994 (chapters 9 and 13 are especially useful because they focus on solids, liquids, and gases and on air); Duit 2009; Fensham, Gunstone, and White 1994; Harvard-Smithsonian Center for Astrophysics (i.e, MOSART); Keeley, Eberle, and Farrin 2005; Keeley, Eberle, and Tugel 2007; Kind 2004; Meaningful Learning Research Group; Olenick 2008; Operation Physics; Osborne and Freyberg 1985; Science Hobbyist; Treagust, Duit, and Fraser 1996; and White and Gunstone 1992. The nature of student misconceptions and their pedagogical implications is a topic for career-long study and professional development.

When Working With Students

Additional experiments can be performed to discuss the misconception about air being "nothing" versus what it actually is: a form of matter that has mass and occupies space (or has volume). One classic demonstration frequently used to challenge the misconception that air is "nothing" is to push two different-shaped, clear, colorless glass containers into an aquarium of colored water so that one container fills with water and the other one is kept upside

down so that air is trapped inside as it is lowered into the water. The trapped air can then be poured "up" into the container of water that is held upside down above the container of air. Understanding gases (and discovery of the gas laws) in the late 1600s and mid-1700s is what helped transform alchemy into chemistry and led to Dalton's atomic theory.

Extensions

If time permits, any of the following demonstration variations may be used to formatively assess whether learners "get" the underlying science concepts and whether they understand the idea that prior knowledge can sometimes be an impediment to new learning:

1. *Empty the Bottle Race.* A challenge can be set up in which the teacher and one student race to be the first to empty one of two identical, glass soft-drink bottles filled with water (or larger jug-shaped containers).

   The bottle can be emptied quickest if it is held at a 45 degree angle and given a rapid twist. This produces the familiar "tornado in a bottle" effect (see Activity #30 in this book) where a "hole" is created for air to enter as water leaves. The glug-glug sound of a bottle emptying is the alternating pattern of water leaving and air entering the bottle (see Internet Connections: Becker Demonstrations/Experiments). Most students will inadvertently demonstrate this phenomenon when they hold the bottle at a greater angle.

2. *Fill That Bottle Race.* Ask learners to explore the relationship between the diameter of a stream of water and the diameter of the mouth of an "empty" bottle that will readily accept the stream of water without resistance from the "trapped" air.

3. *Balloon in a Bottle Version a.* Drill a small hole in one of two otherwise identical 1 or 2 L plastic bottles. Insert identical size but different-color balloons in the open mouth of each bottle so it seems as if the balloons could be blown up inside the bottles. **(See safety warning about use of latex balloons on p. 89.)** Give the bottle that lacks a predrilled hole to a learner and take the "rigged" setup for yourself. Challenge the learner to blow

up the balloon. The internal trapped air will resist compression and the balloon will return to its original, deflated size as soon as the learner stops blowing. The balloon in the teacher's setup, however, will easily enlarge as the trapped air below escapes through the hidden hole. If this hole is subsequently covered by the instructor's finger, the balloon will maintain its enlarged size despite the fact that it is not sealed at its opened end. Alternatively, a piece of tape can be placed over the hole and the bottle-balloon setup can be passed around the room for closer inspection. The sight of a blown-up but unsealed balloon will surprise nearly everyone.

Balloon in Bottle Version b. Boil a small quantity of water in a Pyrex flask until it almost completely vaporizes and then rapidly place an uninflated balloon over the opening of the flask. As the water vapor inside the flask cools and condenses, it leaves a partial vacuum and the atmospheric pressure pushes the balloon tightly into the bottle where it will remain. This "balloon and flask" demonstration is essentially identical to the classic "egg in the bottle" and "can crushing" (or "soda 'pop' can") demonstrations; see Internet Connections: Arbor Scientific for all three).

The idea of a fair test and teaching students to be skeptical of what seem to be identical systems are factors in some discrepant-event, teacher versus student challenges. In such cases, the teacher rigs one of the systems to behave differently; if used properly, such challenges can help students develop observational, inquiry, and thinking skills. This particular demonstration also serves as a visual participatory analogy for how new information is not easily received (nor well retained) if it is in opposition to tenacious misconceptions that are not recognized and "removed."

Internet Connections

- American Educator, Ask a Cognitive Scientist column: *www. danielwillingham.com*. Students remember what they think about: *www.aft.org/pubs-reports/american_educator/summer2003/cogsci. html* and Why don't students like school? Because the mind is not designed for thinking: *www.aft.org/pubs-reports/american_educator/ issues/spring2009/index.htm*

- Arbor Scientific's Cool Stuff Newsletter: *www.arborsci.com/ CoolStuff/Archives3.aspx*. (See Chemistry: Gas laws and pressure and fluids demonstrations.)

- Becker Demonstration/Experiments: Water tornado (empty the jug race): *http://chemmovies.unl.edu/chemistry/beckerdemos/BD010.html*

- Constructivism and Learning Theories: Supplemental readings for teachers: *http://carbon.cudenver.edu/~mryder/itc_data/ constructivism.html* (links to many sites): (1) *http://narst.org/ publications/research/learn.cfm* (constructivism and the learning cycle) (2) *http://narst.org/publications/research/constructivism.cfm* (3) *http://narst.org/publications/research/Metacogn.cfm* (metacognitive strategies) (4) *http://en.wikipedia.org/wiki/Constructivism* (see learning theory and teaching methods) (5) *www.thirteen.org/ edonline/concept2class/constructivism/index_sub5.html*

- University of Virginia Physics Department: Experiments on properties of air (e.g., density): *http://galileo.phys.virginia.edu/ outreach/8thGradeSOL/TrappedInsideFrm.htm*

Answers to Questions in Procedure, step #5

5. The idea of experimental controls and variables is relevant. Neither the color of water nor the height of the boxes (if these were selected to be different) can explain the different outcomes. The volunteer's setup will readily accept the water because the second, unplugged hole allows the air to escape as the water enters (i.e., all matter takes up space or occupies volume). In contrast, the teacher's setup will allow water to enter and air to leave only in alternating spurts. This process is much slower, so the vol-

282

unteer "wins the race." However, the slower setup proves to be additionally discrepant and more interesting in that the water sometimes shoots up and out of the funnel as the trapped air is compressed and rebounds.

Answers to Questions in Debriefing, When Working With Teachers, #1–#4

1. No human has ever come close to "using up" the available trillions of neurons and countless neural connections and networks that are actively constructed (and "rewired") during learning to form the biochemical basis of memory and learning.

2. No, all learners have prior knowledge, just as the seemingly empty bottle contains air.

3. The latter is true (i.e., that new information interacts with prior knowledge), and in some cases, prior misconceptions must be modified, partially dispelled, or even completely expelled before new knowledge can be accommodated and "take hold"—much as the air in the bottle must escape before the water can enter. Learners construct their own somewhat unique understandings from the same external stimulus (provided by a teacher and/or phenomena) as the stimulus interacts with what they already know or believe to be true.

4. Learners, we hope, will have experienced the latter ("kindled fires") in this activity and also will have noted that even hands-off, teacher-directed demonstrations can activate minds to process science concepts. Lack of attention and of mental activity, however, are signs of failed instruction regardless of whether or not students are physically active in hands-on work.

The Stroop Effect: The Persistent Power of Prior Knowledge

image by roccomontoya for iStockphoto

Expected Outcome

The teacher asks learners to state, as fast as they can, the *colors* of a sequence of words that appear in *different colors than the colors named* (e.g., the word *red* printed in *blue*). The first inclination of most people is to *read the words* rather than *naming the colors* in which the words are printed (this inclination is called the Stroop Effect). They can accomplish the color identification task without errors or "miss-takes" only by going very slowly with great concentration.

Science Concepts

The Stroop Effect and Selective Attention Theory: When people who can read look at words printed in color, they both see the color and recognize the word. When those two pieces of information are in conflict—that is, when there is a discrepancy between what color the words represent and what color the words are printed in—most readers will automatically, without consciously thinking, focus on the word's meaning as being more important than font color, despite the stated demand of the assessment task (i.e., to name the color sequence). In this case, prior knowledge of reading interferes with the ability to focus on font color.

In science (and science teaching), the discrepancies between what students expect to "see" and what they actually experience provide teachers with "teachable moments" that can challenge prior conceptions and move students to more refined conceptual understandings. The history of science contains many examples of so-called failed experiments and serendipitous discoveries that led to new insights and fruitful lines of research. Part of the nature of science is the idea that "chance favors the prepared mind" (Louis Pasteur). The prepared mind notices and seeks to explain puzzling discrepancies.

Science Education Concepts

For students of all ages, prior conceptions either can support valid versions of science concepts or they can interfere with learning more scientifically valid versions of science concepts. In this activity, a prior knowledge-assessment task "mismatch" is used as a *visual participatory analogy* for the idea of science misconceptions. In many cases, commonly held misconceptions have historical antecedents that took scientists many years to replace (e.g., the idea of gases as being "nothing" and moving objects "naturally running out of energy"). If teachers intentionally activate and challenge such misconceptions, the misconceptions can be modified or replaced (in part or whole).

Conceptual evolution or the survival-of-the-fittest mental model takes time and requires experiences with multiple discrepant events. Some prior conceptions are especially tenacious (displaying "cognitive inertia") and resistant to change, especially in a quick, cover-the-content approach to teaching. This pedagogical problem is analogous to the short-lasting improvement of painting over a rusty door (rather than properly preparing the surface by removing the layer of old rust before painting).

Materials

Several websites contain both static and interactive versions of the Stroop Effect:

- American Psychological Association: *www.apa.org/science/stroop.html*
- Neuroscience for Kids: *http://faculty.washington.edu/chudler/words.html* (highly interactive)
- NOVA Online Adventure: Shockwave Demonstration version (as well as a static version): *www.pbs.org/wgbh/nova/everest/exposure/stroopintro.html*
- Scientific American Frontiers: *www.pbs.org/saf/1302/teaching/teaching2.htm*

Alternatively, instructors can easily make their own colored transparency or PowerPoint slide by typing various color words (e.g., red, orange, yellow, green, blue, indigo, and violet) in intentionally mismatched colors.

Points to Ponder

The foundation upon which education builds is the equipment of instincts and capacity given by nature apart from training…. The relation between psychology and education is … that action in the world should be guided by the truth about the world; and that any truth about it will directly or indirectly, soon or late, benefit action.

—Edward L. Thorndike, American psychologist (1874–1949) in "The Contribution of Psychology to Education," *The Journal of Educational Psychology*, 1910. Available at *http://psychclassics.yorku. ca/Thorndike/education.htm*

If I had to reduce all of educational psychology to just one principle, I would say this: The most important single factor influencing learning is what the learner already knows. Ascertain this and teach him accordingly.

—David Ausubel, American cognitive psychologist (1918–2008) in *Educational Psychology: A Cognitive View* (1968)

Procedure

Use the interactive websites or type up your own colored-word "tests." Use, for example, overhead transparencies, PowerPoint slides, or T-shirt transfers. Ask for a volunteer to state out loud (as quickly as possible) the sequence of colors in order from left to right, top to bottom, while the rest of the group attempts to do the same task silently. Repeat with another volunteer.

The Stroop Effect readily lends itself to use as a novel "fun"omenon for further inquiry. Run one or more of the following tests in the classroom and/or challenge learners to come up with their own variations of the tests and to use logical argument and skeptical review to explain the empirical results. Ask: What do you think would happen if you…

- tested a younger child who had not yet learned to read (but already knew his or her colors)?
- tested someone who was just learning to read English?
- tested the performance of individuals who had some form of color blindness?
- used the same order of ink colors but used noncolor words (e.g., *school, books,* or *science*)?
- used unrecognizable foreign words or nonsense words? [no cognitive conflict occurs]
- tested if accuracy is affected if the colors of the first couple of words are the color words themselves—e.g., the word *red* appears in red; the word *blue* appears in blue—before the mismatched examples?
- turned the words upside down so that the mismatched words were not easily readable?

Debriefing

When Working With Teachers and Students

Both teachers and students need to directly experience the power of prior cognitive conceptions and to become aware of the mental effort that is needed to identify and confront our own misconceptions. The skills of disciplined scientific inquiry have utility in day-to-day life as well as in a science classroom. The metaphor of *conceptual* survival of the fittest helps us to understand the cognitive growth of any individual and the history of science. In the classroom, different ideas and theories "compete for survival" in the minds of individual students. Unfortunately, the correct science theory often loses out to the erroneous prior conceptions (or misconceptions) the students bring to class.

The metaphor of conceptual survival also points to the need for disciplined inquiry. Simple interpretations of uncontrolled everyday natural phenomena often lead to incorrect personal, "scientific" theories. Ausubel's quote suggests that when planning curriculum, instruction, and assessment, teachers need to assess what students already know (some of which is "not so" in terms of scientifically valid theories). Novel discrepant-event activities are especially useful as diagnostic and instructional tools for activating prior conceptions and catalyzing cognitive processing, two strategies that lie at the heart of constructivist science teaching.

Extensions

1. *Scientific Snooping on the Stroop Effect.* See the websites in the Materials section, as well as Bower 1992; MacLeod 1991; and Stroop 1935.

2. *More Modeling of Mental Blinder.* Repeat the Stroop Test with volunteers who are wearing color-tinted sunglasses. They will find that words printed in the complementary color of the lenses (i.e., red-green, orange-blue, and yellow-violet) become black. For example, orange-tinted blue blocker sunglasses absorb blue light before it reaches the eye, so words printed in blue will not be visible. Tinted glasses can be used as another analogy for how our prior conceptions can influence or "color" how we subjectively perceive, interpret, and reconstruct reality. Creative insights and new technologies often advance scientific understanding by removing conceptual blinders or human sensory limitations. See also Activity #26 in this book.

3. For connections to serendipitous scientific discoveries and to explore the research on misconceptions in science, see Internet Connections.

Internet Connections

(These sites are in addition to those listed above in the Materials and Points to Ponder sections.)

- Access Excellence: Discovery, chance and the scientific method: *www.accessexcellence.org/AE/AEC/CC/chance.html*

- Duit, R. Free download (March 2009 update): About 8,400 entries related to "misconceptions": Bibliography of STCSE (Students' and Teachers' Conceptions and Science Education). *www.ipn.uni-kiel.de/aktuell/stcse/stcse.html*

- Kind, V. 2004. Beyond appearance: Students' misconceptions about basic chemical ideas: *www.rsc.org/education/teachers/learnnet/pdf/LearnNet/rsc/miscon.pdf* (book; 84 pages)

- MOSART: Misconception oriented standards-based assessment resource: *www.cfa.harvard.edu/sed/projects/mosart.html.*

- Operation Physics: Children's misconceptions about science:
  www.amasci.com/miscon/opphys.html

- Wikipedia: (1) Attention: *http://en.wikipedia.org/wiki/Attention*
  (2) Color blindness: *http://en.wikipedia.org/wiki/Color_blindness*
  (3) Constructivism: *http://en.wikipedia.org/wiki/Constructivism_
  (learning_theory)* (4) List of cognitive biases: *http://en.wikipedia.
  org/wiki/List_of_cognitive_biases* (5) Serendipity (links to many
  examples): *http://en.wikipedia.org/wiki/Serendipity* (6) Stroop
  effect: *http://en.wikipedia.org/wiki/Stroop_effect*

Activity 29

Rattlebacks: Prior Beliefs and Models for Eggciting Science

Expected Outcome

A translucent, half-ellipsoid-shaped, molded acrylic polystyrene object —known as a "rattleback"—is placed on an overhead projector or under a document camera and is observed to spin freely if pushed in a counterclockwise direction. When spun in a clockwise direction, it slows down quickly, starts to rock in a vertical direction, and reverses to its "natural" counterclockwise rotation. Similarly if the rattleback is pushed down on either end, the resulting rocking motion changes to rotation in the "preferred" direction of movement.

Science Concepts

The science concepts addressed with this surprisingly complex, yet simple toy include the nature of science (i.e., evolution of theories); center of gravity; friction; rotational dynamics; stability; and inertia and conservation of energy.

There are two basic designs for rattlebacks. One features a symmetrical hull with offset weighting at one end; the other has a nonsymmetrical hull with a skewed alignment to the keel but uniform weighting. The sources in the Materials section sell the latter type (which are designed to "prefer" a counterclockwise motion). The original, ancient prototypes for modern day rattlebacks (or celts or "space pets") were discovered about a century ago by archaeologists studying prehistoric stone axes and adzes that exhibited this behavior. The physics and mathematics of their motion is quite complex. Therefore, the "toy" is best used as a source of curiosity and as a *visual participatory analogy* for human resistance to giving up prior cognitive commitments at the individual, science discipline, and societal levels. For high school students and their teachers, the toy can also be used as a visual metaphor for Thomas Kuhn's (1962) ideas of "normal" and "revolutionary" science and paradigms. The Extension activity with fresh eggs and hard-boiled eggs is a similar black box system that can be "dissected" and explored with models. Eggs can also serve as a not-to-scale model of the lithospheric layers of the Earth.

Science Education Concepts

Prior conceptions (and cognitive biases) are tenacious and resistant to change even in the face of new information that undermines their validity. Many prior conceptions can be used as solid building blocks for constructing new or refined scientific conceptions, but others are misconceptions that need to be challenged. Trying to overlay new science concepts on top of unexposed, unexamined misconceptions is analogous to putting fresh paint over a rusting door. In the short term, things might look good, but a little later the new paint flakes off, revealing the previous unaltered rusty door. For long-term effectiveness, the rust needs to be removed before painting rather than merely painted over for a short-term effect. So, too, must misconceptions be exposed and examined. This activity also models how science can be considered a form of planned and purposeful playfulness.

Safety Notes

1. Remind students not to eat food used in an activity or in a laboratory.
2. Always clean up clay when it is wet. Dry clay contains silica, which is a health hazard when airborne.

Materials

- Science Supply Sources for Rattlebacks, Celts, and Space Pets:
  - Arbor Scientific. *http://arborsci.co.* 800-367-6695. #P2-2120. 2 celts (9.8 cm long) for $3.95.
  - Educational Innovations. *www.teachersource.com.* 888-912-7474. #SS-300. 2 for $3.25.
  - Edmund Scientific. *http://scientificsonline.com.* 800-728-6999. #3039089. 1 Space Pet for $1.95.
- *Optional* for Extension #2a and #2b:
  - Dimes (also needed for Procedure, step #2b)
  - Tape (also needed for Procedure, step #2b) Type of tape doesn't matter—either transparent or masking tape is fine.
  - 1 or more fresh eggs
  - 1 or more hard-boiled eggs
  - Plastic, hollow eggs (typically sold at Easter time)
  - Clay and water (to be used to construct models—from the plastic eggs—of the two kinds of eggs)

Points to Ponder

The human understanding is no dry light, but receives infusion from the will and affections; whence proceed sciences that may be called "science" as one would... what a man had rather were true he more readily believes. Therefore he rejects difficult things from impatience of research; ... the deeper things of nature, from superstition; the light of experience, from arrogance and pride; things not commonly believed, out of deference to the opinion of the vulgar. Numberless in short are the ways, and sometimes imperceptible, in which the affections color and infect the understanding.... Men have been kept back as by a kind of enchantment from progress in the sciences by reverence for antiquity, by the authority of men counted great in philosophy, and then by general consent.

—Francis Bacon, English essayist and philosopher (1561–1626) in *The New Organon* (1620)

Procedure

(For answers to questions in step #1, see p. 299.)

1. Introduce the activity by projecting or reading the Francis Bacon quote and briefly discussing the positive and negative outcomes of "cognitive conservatism" (i.e., continuing to believe what we have always thought was true, even in the face of evidence that it is not). Ask: Is everything printed in current science textbooks "true"? What past scientific "truths" proposed by scientific giants or superstars have later been proven either false or of limited applicability relative to more universal natural principles?

2a. *Teacher Demonstration Version:* Place the rattleback on an overhead projector or under a document camera and rotate it in the counterclockwise direction. No surprise will be noted. Suggest that much "normal science" proceeds in this fashion, with new ideas nicely fitting into the broader context of previous theories. Next, give it a push in the clockwise direction and it will be observed to resist that motion by first rocking up and down and then reversing to its "preferred" counterclockwise direction. Elicit ideas to account for this unusual behavior.

2b. *Student Hands-On Exploration Version:* If you have at least eight rattlebacks, instead of starting with a demonstration distribute the toys with the flat side down and challenge students to play and discover what they can do with them. Some students may discover their optical properties (which vary depending on which side you look through); others may discover their rotational properties. If no group discovers their rotational "bias," suggest they explore this phenomenon.

   Also challenge students to see if they can make the rattlebacks spin by pushing in a downward direction. Provide students with dimes or heavier coins and tape to see if they can modify the rattlebacks so that they can spin freely in either direction (without reversing). This latter test helps confirm that the unique behavior of the rattleback has something to do with center of gravity (or mass). You may want to sacrifice one toy for bi-section (using a hacksaw to cut across the width) to compare the masses of the two halves. Consider showing the segment of the NASA Toys in Space online video that features a rattleback's behavior in a microgravity environment (see Internet Connections: Toys in Space II).

Debriefing

When Working With Teachers and With Students

The purpose of this particular activity (as outlined in the Procedure section) is less about teaching particular science concepts than it is about teaching that part of the nature of science is to challenge status quo thinking. It is in the nature of human beings, however, to resist changing their prior conceptions, even when faced with evidence that raises questions about the validity of these prior beliefs.

Both students and teachers need to learn to critically examine both new information and their prior beliefs. Science should "make sense" rather than just be blindly accepted on the basis of the authority of the teacher or textbook. Teachers can use the Internet Connections, below, to explore published research on science misconceptions to stimulate their interest in using preinstructional assessments and curriculum-embedded formative assessment.

Extensions

1. *Eggs-perimentation and Model Making.* An alternative discrepant event involving a rotational system would be to contrast the spinning behavior of fresh eggs and hard-boiled eggs.

   a. *Teacher Demonstration.* Lay the eggs horizontally on their sides on an overhead projector or under a document camera. Use your thumb and index finger to spin the eggs on their sides (works equally well in either direction). The fresh egg is difficult to get moving and rotates slowly (because of internal friction), but if it is quickly touch-stopped with an index finger and then released, it will resume rotating in the same direction. Conversely, the hard-boiled egg can spin quite rapidly, but when touch-stopped it will not resume its motion (i.e., it does not have internal parts that are disconnected from the shell that can keep moving when the shell is stopped). This serves as (a) an easy way to distinguish unmarked eggs, (b) a model for the Earth's lithospheric layers and overall rotation, and (c) a nice black box system to study. (*Note:* If the hard-boiled egg is spun quickly enough, it will rise up on end and spin like a top.)

b. *Student Hands-On Exploration Activity.* Ask the learners to brainstorm a variety of nondestructive means of distinguishing between a fresh egg and a hard-boiled egg [by shaking or spinning]. After students examine the two kinds of unmarked eggs, challenge them to construct working models of the insides of fresh eggs and of hard-boiled eggs using plastic, hollow "Easter eggs" to see if they can replicate the different spinning behaviors (using clay and water as fillers). Also, if a fresh egg is shaken vigorously, it can be made to sit on end because the previously suspended and anchored yolk will drop inside the egg, lowering the egg's center of gravity. Other egg-citing explorations can be found in the book *Teaching With Eggs* (Devito 1982).

2. Books that explore the history and conceptual origins of "wrong" ideas in science include Grant 2006, Wolpert 1993, and Youngson 1998. The books by Grant and Youngson contain numerous examples of scientific "miss-takes" from the early history of science up to the cold fusion affair.

Internet Connections

- Arbor Scientific's Cool stuff newsletter: *www.arborsci.com/ CoolStuff/Archives3.aspx.* (See Force and motion: Egg pizza inertia, Egg spin, and related demonstrations.)

- Encyclopedia of Science: Chaper 14: *www.daviddarling.info/ encyclopedia/C/celt.html*

- Toys in Space II: *www.nasa.gov.* Search for Toys in Space II video resource guide. Online video (38 min.; includes celts): *http:// quest.nasa.gov/content/rafiles/space/toys.rm*

- University of Iowa Physics and Astronomy Lecture Demonstrations (video clips): Mechanics: *http://faraday.physics.uiowa.edu* (Rotational stability: 1Q60.36—Spinning eggs).

- Wikipedia: Rattleback: *http://en.wikipedia.org/wiki/Rattleback*

Research on Misconceptions in Science:

- Duit, R. 2009 update (about 8,400 entries). Bibliography–STCSE (Students' and teachers' conceptions and science education): *www.ipn.uni-kiel.de/aktuell/stcse/stcse.html*

- Kind, V. Beyond appearance: Students' misconceptions about basic chemical ideas: *www.rsc.org/education/teachers/learnnet/pdf/ LearnNet/rsc/miscon.pdf*

- Meaningful Learning Research Group: Misconceptions conference proceedings: *www2.ucsc.edu/mlrg/mlrgarticles.html*

- Science Hobbyist: Amateur science: Science myths in K–6 textbooks and popular culture: *www.amasci.com/miscon/miscon.html*

- Wikipedia: List of cognitive biases: *http://en.wikipedia.org/wiki/ List_of_cognitive_biases*

Answers to Questions in Procedure, step #1

1. Numerous historical examples exist of long periods of stasis in science that are infrequently punctuated by revolutionary ideas. Consider for example the historical "evolution" of successive atomic models in chemistry: Aristotelian→Newtonian→ Einsteinian physics; the theory of biological evolution; and beliefs about change and uniformity over Earth's geologic history. See Extensions, step #2 for resources about scientific "miss-takes."

Tornado in a Bottle: The Vortex of Teaching and Learning

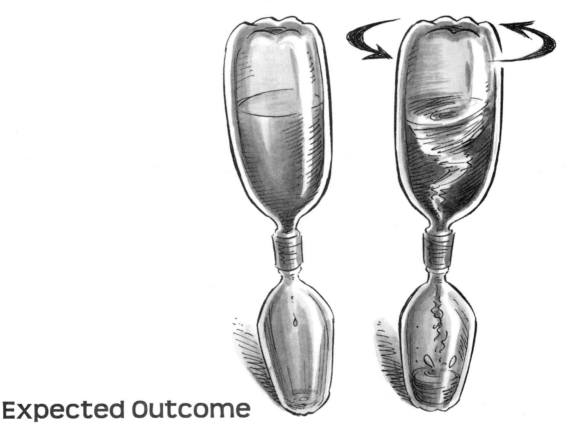

Expected Outcome

Two 2 L plastic soda bottles are connected at their mouths. Colored water from the upper bottle falls into the lower bottle quickly *only after* the two bottle system is given a twist to create a spiraling, funnel-shaped vortex.

Science Concepts

Gases (including colorless, invisible ones such as air) are a form of matter that occupy space (i.e., have volume) and have a lower density than liquid water. Also, some aspects of the nature of vortex systems in waterspouts, tornadoes, hurricanes, weather systems, galaxies, and black holes can be visually modeled. At the level of high school physics, angular momentum can be discussed.

Science Education Concepts

Science instruction catalyzes students' cognitive construction by engaging students with new experiences in which their prior knowledge is activated ("shaken up") to see what is still valid and what needs to be extended, modified, or replaced ("to allow the new concepts in"). However, knowledge is not like a fluid substance held by a teacher that can be transferred into the empty vessel of students' minds without any mental action by the students. This *visual participatory analogy* can be used to challenge the empty vessel idea and other "unquestioned answers" about interactive teaching and minds-on learning (see discussion in Debriefing, When Working With Teachers, p. 304).

Materials

- 2, 2 L plastic soda bottles. The lower bottle should be one-half to two-thirds filled with colored water and then connected to the upper bottle with a *vortex bottle connector*. These connectors are available at science museums, novelty shops, and science education supply companies such as the following:
  - Arbor Scientific. *http://arborsci.co.* 800-367-6695. #P1-1120. $1.50/one connector
  - Educational Innovations. *www.teachersource.com.* 888-912-7474. #SS-1. $1.49
- Homemade Connectors
  - Construct a connector by simultaneously drilling 0.8–0.9 cm holes through the middle of two plastic bottle caps. Glue and water-seal the top of the two caps together, and wrap a piece of duct tape around the homemade connector.

Safety Note

Students and teacher should wear safety glasses or goggles during this activity.

- (Even quicker!) Place a washer with a 3/8 in. (0.9 cm) hole on top of the water-filled bottle. Put the air-filled bottle on top and wrap duct tape tightly around the two mouths. This connector does not allow for the easy separation and reconnection of either the commercial version or the two-cap homemade version, but otherwise it works as well. Also, you may wish to drill a hole near the top (on the side) of one bottle; the hole can be covered with tape or left open as desired.

Points to Ponder

Nature produces those things that, being continually moved by a certain principle contained in themselves, arrive at a certain end…. Nature abhors a vacuum.

—Aristotle, Greek philosopher (384–322 BC)

There are three distinctions in the kinds of bodies, or three states, which have more especially claimed the attention of philosophical chemists; namely those which are marked by the terms elastic fluids, liquids, and solids.

—John Dalton, English physicist-chemist (1766–1844)

Procedure

(For answers to questions in steps #1–#5, see pp. 306–307.)

1. Present two capped 2 L bottles, one containing only air and the other one-half to two-thirds filled with colored water. Ask learners to describe what they see.

2. Use the vortex bottle connector to connect the two bottles, with the colored water bottle on the bottom. Ask learners to *predict:* What will happen if you invert the two bottle system, placing the colored water bottle on top? Why?

3. Invert the bottle and ask the learners to *observe* the behavior of the liquid. Ask the learners to *explain* the "reluctance" of water to transfer. Aristotle would argue that "nature abhors a vacuum." Is this a good explanation? Alternatively, consider whether the lower bottle is really "empty," and if not, how its contents might be a factor. How might we speed up the exchange of fluids between the top bottle of water and the bottom bottle of air? Elicit multiple hypotheses.

4. Give the top, water-filled bottle a rapid twist, and a vortex will form that allows the rapid exchange of liquid water and gaseous air. Would this two bottle system work the same way in a weightless or microgravity environment? Does water "want to" fall and air "want to" rise in keeping with Aristotle's belief that they are "moved by a certain principle contained in themselves [to] arrive at a certain end"?

5. How does this system help explain why we must punch two holes in a juice can to efficiently pour juice out of the can? How does the shape of the pull-tab hole in soda cans replicate a two-hole system?

Debriefing

When Working With Teachers

Critique the use of this two bottle system as a visual participatory analogy for learning, with the colored water representing science concepts. Can knowledge be "transferred" from the teacher or textbook to a student who passively receives it? Are students' minds empty vessels prior to instruction? What are other limitations of this analogy? How can appropriate instructional "twists" create cognitive interaction or disequilibrium between newly introduced concepts and students' prior knowledge?

These questions are designed to challenge some of the unquestioned answers or tacit assumptions and implicit learning theories that teachers (and students) may unconsciously hold. Knowledge is not a transferable substance; teachers are not emptied of knowledge as they teach. Learning is not something that teachers can do to, or for, passive students. New knowledge rarely completely replaces previous knowledge even in the case of misconceptions. Discrepant-event activities are especially useful in activating and shaking up students' prior knowledge and catalyzing learning. Unlike the two bottle system, however, students must do the hard, minds-on work of cognitive (re)constructions. The Internet Connections section, below, provides several links to short synopses of constructivist learning theory for interested teachers.

When Working With Students

Discuss the historic importance of the recognition of colorless, invisible gases (Dalton's "elastic fluids") as a form of matter that must be measured, weighed, and otherwise accounted for in scientific experiments. The systematic study of gases during the 1700s played a major role in the development of both the law of conservation of matter (by Lavoisier) and Dalton's atomic theory. This activity is a useful Engage-phase activity in a unit focused on establishing the empirical evidence and logical arguments for Daltonian atoms and the kinetic molecular theory that should precede any discussion of modern atomic theory.

Extensions

1. *Viscosity Variations.* Test liquids of different viscosities—e.g., water, olive oil, corn syrup, and honey—in terms of how fast they run in side-by-side tornado tube races. Bottles that are 1 L or smaller can be used in place of the 2 L bottles.

2. Educational Innovations' Fountain Connection (#SS-10 @ $3.95) uses a two-hole system and hard plastic straw to produce a modern version of the fountain that was created 2,000 years ago by the Greek inventor, mathematician, and physicist Hero. Also available is the Vortex (#SS-400 @ $26.95), a plastic vortex for studying the physics of rotational motion using coins.

3. See the book by Steve Spangler, *Taming the Tornado Tube: 50 Weird and Wacky Things You Can Do With a Tornado Tube* (1995). In addition to the short experiments in this book, the author describes the tornado tube's history (a result of a failed hourglass design by a young boy in 1964 that he eventually patented and financed 25 years later!) and a little bit of the science behind tornadoes and hurricanes.

Internet Connections

- Arbor Scientific's Cool Stuff Newsletter: *www.arborsci.com/CoolStuff/Archives3.aspx*. (See Pressure and fluids demonstrations.)

- Constructivism and Learning Theories: Supplemental readings for teachers: (1) *http://carbon.cudenver.edu/~mryder/itc_data/constructivism.html* (2) *http://narst.org/publications/research/learn.cfm* (3) *http://narst.org/publications/research/constructivism.cfm* (4) *http://narst.org/publications/research/Metacogn.cfm* (5) *http://en.wikipedia.org/wiki/Constructivism_(learning_theory)*

- Exploratorium Snackbook: Vortex (tornado-in-a-bottle demonstration): *www.exploratorium.edu/snacks/vortex/index.html*

- Surfing Scientist: Vortex rings: *www.abc.net.au/science/surfingscientist/toroidalvortex.htm*

- University of Iowa's Fluids Laboratory Image Gallery: Vortices: *www.engineering.uiowa.edu/fluidslab/gallery/vortex.html*

- Wikipedia: Vortex: *http://en.wikipedia.org/wiki/Vortex* and Viscosity: *http://en.wikipedia.org/wiki/Viscosity#Viscosity_of_various_materials*

Answers to Questions in Procedure, steps #1–#5

1. Some learners will probably say that one bottle is "empty," as opposed to being filled with a mixture of gases that we call air; do not correct them at this time.

2. Some learners will say that the colored water transfers or falls into the lower bottle due to gravity.

3. If the two bottle system is inverted slowly with minimal shaking, the colored water will either remain on top or fall very slowly into the lower bottle in alternating spurts and stops. Water resists flowing into the lower "empty" bottle because that bottle is really "filled" with air, not because "nature abhors a vacuum" (a vacuum might be envisioned developing in the upper bottle as water falls into the lower one). The lower density air must be allowed to transfer into the upper, water-filled bottle to make room

for the falling water. The exchange of the two fluids is restricted by the narrow opening in the connector, by the presence of water, and by the surface tension of the water at the small opening. If the transfer occurs at all, it will do so in fits and starts because the air below becomes pressurized and is forced upward, releases pressure, and becomes re-pressurized by the addition of more water. The air will transfer quickly if the system is given a rapid circular twist, producing a vortex of swirling liquid.

Other options to explore include (a) drilling a hole near the top side of the air bottle and covering it with a removable piece of tape to test whether opening the hole makes a difference and (b) drilling a hole in the bottom of the water bottle. (*Note:* The tornado tube is a variation of Activity #27 in this book.)

4. No, the transfer is driven by gravity; fluids do not "fall down" and are not "pushed up" as there is no "up" or "down" in the absence of gravity. Gravitational forces always involve an interaction between at least two masses; water (or air) in this system does not move downward (or upward) by itself but rather in relationship to the gravitational attraction with the Earth.

5. Two holes allow air to enter the can as the liquid leaves, which allows for a faster transfer. The shape of the pull-tab hole at the top of a soft drink can effectively makes the can a two-hole system when it is brought up against the drinker's lips.

Activity 31

Floating and Sinking: Raising FUNdaMENTAL Questions

RAISINS

Expected Outcome

Raisins are observed to alternately rise (float) and fall (sink) in a cup of clear, colorless, carbonated soft drink. This cycle continues until the drink goes "flat."

Science Concepts

This activity can be used to explore density and buoyancy; the solubility of gases in liquids under different temperatures and/or pressures; diffusion and osmosis; and nucleation sites with applications across physics, chemistry, biology, and Earth science. It also playfully challenges the notion of a single, monolithic, follow-the-dots scientific method.

Science Education Concepts

Learning science is FUNdaMENTAL—that is, it is a combination of hands-on FUN and minds-on MENTAL activity. It builds on students' prior knowledge that has been acquired both from formal instruction (including carefully designed experiments) and from daily life experiences. Even seemingly simple materials and phenomena tend to reveal hidden delights to the meticulous observer. Direct experience with the simple joys (and frustrations) of hands-on explorations with everyday phenomena is essential in helping teachers learn *how* to learn—and also learn how to teach science in more effective, research-informed ways.

This activity is a visual participatory analogy for, and a concrete example of, interactive learning experiences that challenge learners to raise questions and "question answers" about science and challenge teachers to learn more about the science and art of teaching. As such, it models the inquiry-based nature of science orientation promoted by the National Science Education Standards and the Benchmarks for Science Literacy (NRC 1996; AAAS 1993). The 7 Ps of Scientific Discovery (see the Procedure section, p. 312) is a lighthearted attempt to poke a little fun at the idea of a rigid sequence of steps in a "scientific method." Teachers and students should come to science with an open-ended sense of purposeful playfulness. As teachers re-discover their childhood curiosity in questioning "simple" things, they will be more inclined to have their science classes catalyze student construction of rich, meaningful conceptual networks associated with deep, lasting understanding. This approach stands in contrast to science classes in which surface-level understanding of memorized content and procedures is emphasized, regurgitated mindlessly for low-level tests, and quickly forgotten.

Materials

- 1 tablespoon of raisins for each group of two or three learners
- Enough 7-Up (or Sprite or other colorless, carbonated soft drink) to nearly fill a cup for each group
- Clear colorless plastic cups
- Hand lenses or 30× handheld microscopes
- *Optional Items for Biology or Life Science Connections* (see Procedure): four-ball tennis can or small-diameter glass flower vase; Mountain Dew
- *Optional Items to Experiment With:* graduated cylinders, balances, non-mercury thermometer, stopwatches, and clay or Play Dough
- *Optional Background Music:* Songs such as "Heard It Through the Grapevine," "You Raise Me Up," and "Up, Up, and Away" ("[in] my beautiful, my beautiful balloon")

Safety Notes

1. Students and teacher should wear indirectly vented chemical splash goggles during this activity.

2. Remind students not to drink soda or eat food used in an activity or a laboratory.

3. Always clean up clay when it is wet. Dry clay contains silica, which is a health hazard when airborne.

Points to Ponder

The formulation of a problem is often more essential than its solution, which may be merely a matter of mathematical or experimental skill. To raise new questions, new possibilities, to regard old problems from a new angle, requires creative imagination and marks real advances in science.

—Albert Einstein, German-American physicist (1879–1955) in *The Evolution of Physics* (1938)

We begin with the hypothesis that any subject can be taught effectively in some intellectually honest form to any child at any stage of development. It is a bold hypothesis and an essential one in thinking about the nature of curriculum.

—Jerome Bruner, American cognitive psychologist (1915–) in *The Process of Education* (1960)

Procedure

(Answers to questions in steps #1, #2, and #4–#7 are found on pp. 317–320.)

Use guided inquiry to lead the learners through the following seven steps, taking care to avoid direct teaching ("telling the answers") or providing answers to their questions or confirmations of their ideas until the conclusion of the activity.

An alternative biology or life science connection that can be done before undertaking the seven steps: Tell your students that biologists are currently testing various kinds of genetically modified maggots (aka raisins) as potential natural recyclers for biohazardous materials such as human urine (aka yellow-colored Mountain Dew). After introducing the topic, reveal a four-ball tennis can or small-diameter glass flower vase that is filled with the "biohazardous material" (Mountain Dew). Drop a few of the maggots (raisins) into the liquid and watch for the students' reactions as the maggots appear to come alive in the liquid. For extra effect, pop a few of the raisins into your mouth. By now most of the students will be aware of the ruse. The concepts of diffusion, osmosis, and floating and sinking are especially relevant to biological organisms even though the maggots are really raisins. Furthermore, biologists today are genetically altering bacteria for a variety of purposes, including "eating up" oil spills.

Proceed to the 7 Ps of Scientific Discovery. The much more dramatic, large-scale teacher demonstration in the first Extension can be done either before or after the hands-on exploration with raisins.

1. *Perceive Phenomenon.* Separately examine a raisin and a small quantity of fresh 7-Up (or other colorless, carbonated soft drink in a plastic cup). (*Note:* Re-cap the opened bottle to preserve carbonation for later use.) Consider the physical states, surface textures, and likely compositions and origins of the two substances. Describe what you can directly observe (as connected to and separate from your prior knowledge and inferences). What senses can you use to study the two substances? What scientific instruments could extend the limits or increase the precision of your senses? Give each group of two or three learners a hand lens or a 30× handheld microscope for a closer look at the raisins.

2. **Ponder** *the possible* outcomes of what might occur if raisins are dropped into the glass of 7-Up. Consider the physical, chemical, and/or biological processes that are likely to operate in this "simple system." A quick round of think-write-pair-share will help to encourage both individual and small-group effort. The teacher's role is to encourage wide participation in brainstorming in the small groups. Avoid premature closure on the "single right answer"; do not attempt to confirm or deny the validity of any ideas. Rather, simply record the learners' ideas so all can see and reflect on them.

3. In their groups, learners **predict** (based on their observations and prior knowledge) which of the listed possible events are most likely to happen when the raisins are dropped in the 7-Up. Learners should give the *experiential evidence (or empirical data)* and *logical arguments* for their predictions. Then they should challenge one another's ideas *with skeptical review*. This discussion can help learners carry out the following step.

4. Learners **plan,** *perform, and observe* an experiment. Ask: What variables should or can you control? Have learners drop one or more raisins into *freshly poured* 7-Up and record their results, especially noting any discrepancies relative to their predictions. For extra fun, play some background music (see Materials).

5. Learners **postulate** *a refined theory* that accounts for the actual observations that forms the basis for next-step investigations. Tell learners that scientists often build *models* to explain specific aspects of a given phenomenon. Assuming that the learners see that the wrinkled surface of the raisin appears to be a key factor in capturing and releasing the gas bubbles that cause the raisins to periodically rise and then fall, challenge them to construct a model that could be used to further test the significance of a dimpled surface area and the capture and release of gas bubbles. Provide learners with modeling clay and pins with small heads to construct their models.

6. **Publish** *the results* and engage in professional *skeptical review* that leads to another round of hypothesis formation and testing. Ask: What related experiments would you like to explore next? (*Note:* Preparing laboratory reports for others to read and critically

review is as important for learners as it is for practicing scientists. Inquiry-based science courses provide opportunities for learners to practice writing extended, logical arguments—and rewriting them as needed. Writing is a powerful means of clarifying thinking and exposing it for critical feedback.)

7. *Practical* applications often result when scientists learn more about the form and function of natural systems. Challenge the learners to think of other real-world systems that can change their buoyancy to either rise and float or fall and sink in their fluid (i.e., liquid or gaseous) environments.

Debriefing

When Working With Teachers

Challenge teachers to consider how this activity is a model for a simple, inquiry-oriented, hands-on exploration and how they could modify it for use across a range of grade levels. Emphasize that the formal introduction of scientific terminology should be delayed until after learners have one or more experiences that create a need-to-know. Also, use the 7 Ps of Scientific Discovery to question the idea of a formal scientific method while still acknowledging the need for a logical plan to answer questions and explore phenomena. This idea of a logical plan also applies to constructivist teaching models such as the BSCS 5E Teaching Cycle (Engage, Explore, Explain, Elaborate, Evaluate) to promote a "FUNomena first, facts follow" approach to unit planning. Also, interested teachers can explore the research on student misconceptions related to gases and density-floating/sinking (e.g., see Driver et al. 1994; see chapters 8, 9, 12, and 13 as starting points).

Discuss how even "simple" systems (e.g., raisin + soft drink) can display relatively complex behaviors that involve underlying scientific principles that cut across the artificial boundaries of multiple science disciplines. Use the two quotes to discuss how problem defining and explaining in science can occur at multiple levels of analysis in a manner analogous to layers of an onion. As such, many inquiry activities are worth reconsidering at various points across the K–12 spectrum and in different science disciplines.

An analogy can be drawn between floating and sinking as a system-level, interactive phenomena (e.g., the relative density of a *particular* solid in a *given* liquid) and the "rising success" or "sinking failure" of a particular student in a given learning environment. Differentiated curriculum-instruction-assessment plans, as reflected in quality 5E teaching cycles, are designed to provide support to help all students "rise" to their full potentials. Without minimizing the primary importance of student learning efforts, discuss to what extent student failures to learn science are due to the "conceptual density" of the learning environment rather than the perceived "mental density" of students. The former explanation leads us to consider how we can latch onto students' unique prior knowledge with pedagogical hooks that help them pull themselves toward higher levels of understanding and competence. In contrast, the latter explanation allows us to ignore students' unique "surface features" and let them "sink or swim" on their own.

When Working With Students

(See the answers to the questions in the Procedure section on pages 317–320.)

Extensions

1. *Mentos Geyser–Diet Coke Eruption* (Teacher Demonstration). Instructions and a video demonstration of the rapid de-gasification of carbonated soda on the large surface area to volume ratio of the candy can be found at *www.stevespanglerscience.com/experiment/00000109*. A roll or box of Mentos (candy mints) and a 2 L bottle of diet soda are needed (either diet or regular soda will work for this experiment, but diet soda is less sticky when you're cleaning it up). This is a more dramatic version of the release of bubbles caused by nucleation of a supersaturated dissolved gas on a solid with a large surface area. (*Safety Note:* Given the force, speed, and height [up to 25 ft.!] of the eruption and the resulting mess, do this demonstration outside and make sure that the bottle is secure so that it cannot topple over. Have students wear appropriate eye protection and stand several yards back.)

If desired, any number of variables can be changed (e.g., the type of hard candy or the carbonated drink used, the temperature of the soft drink, whether or not the candy is broken up, and so forth). Although the demonstration can be done without it, the Geyser Tube (Item #WGEY-500 for $4.95 from *www. stevespanglerscience.com/product/2072*) is a useful means to produce a maximum height effect. See also Arbor Scientific's Cool Stuff Newsletter: Chemistry: Mint Magic (Mentos and Diet Coke): *www.arborsci.com/CoolStuff/Archives3.aspx*.

2. *Fruits Frequently Float but Some Don't* (Student Hands-On Explorations). Challenge students to predict, test, and measure the relative and actual densities of an array of different types of fruit. "Sinkers" could include a raisin, blueberry, seedless grape, grape tomato, pear, apricot, and green plum. "Floaters" could include a peach, banana, lime, nectarine, regular tomato, orange, clementine, lemon, apple, and orange pepper. Some variation will be discovered within different varieties of the same basic kind of fruit. Also, most fruits do not have a uniform density and de-skinned or cut-up pieces may behave differently than intact, whole fruit. Students should note that both mass and volume are important in determining whether a fruit floats in a given liquid. Also, fruits that sink in freshwater may float in saturated saltwater or sugar-water solutions because these solutions are denser than freshwater. Conversely, fruits that float in water may sink in less-dense liquids such as alcohol or cooking oil.

3. A large number of discrepant density demonstrations and engaging experiments have been published (e.g., O'Brien, Stannard, and Telesca 1994). See also Activity #32 in this book and the Internet Connections, below.

Internet Connections

- American Educator (see also *www.danielwillingham.com* for other articles and videos): Critical thinking: Why is it so hard to teach? (feature article): *www.aft.org/pubs-reports/american_educator/issues/summer07/index.htm* and What is developmentally appropriate practice? (pdf download from Ask a Cognitive Scientist): *www.aft.org/pubs-reports/american_educator/issues/summer08/index.htm*

- Arbor Scientific's Cool Stuff Newsletter: *www.arborsci.com/CoolStuff/Archives3.aspx* (See Chemistry: Gas laws, for a variety of demonstrations.) and Diving insect regulates buoyancy (World first!): *www.adelaide.edu.au/adelaidean/issues/12121/news12122.html*

- Concord Consortium (free downloadable simulations): (1) Diffusion: *http://concord.org/modeler (2) Osmosis: www.concord.org/~btinker/workbench_web/models/osmosis.swf (3) http://mw.concord.org/modeler/molecular/html*

- Diving Insect Regulates Buoyancy: *www.adelaide.edu.au/adelaidean/issues/12121/news12122.html*

- Java Applets for Physics: Buoyant force in liquids: *www.walter-fendt.de/ph14e/index.html*

- Purdue University, Chemical Education: Quick Time movies of five density demonstrations: *http://chemed.chem.purdue.edu/demos/index.html*

- University of Iowa Physics and Astronomy Lecture Demonstrations (video clips): Heat and fluids: *http://faraday.physics.uiowa.edu* (five demonstrations on density and buoyancy)

- Wikipedia: (1) Buoyancy: *http://en.wikipedia.org/wiki/Buoyancy* (including applications) (2) Density: *http://en.wikipedia.org/wiki/Density* (3) Henry's law: http://en.wikipedia.org/wiki/Henry%27s_Law (4) Priestley, Joseph: *http://en.wikipedia.org/wiki/Joseph_Priestley* (invented carbonated water) (5) Nucleation: *http://en.wikipedia.org/wiki/Nucleation* (6) *http://en.wikipedia.org/wiki/Diet_Coke_and_Mentos_eruption* (7) Raisin: *http://en.wikipedia.org/wiki/Raisin*

Answers to Questions in Procedure, steps #1–#2 and #4–#7

1. Raisins are partially dehydrated, naturally sweet grapes that have a fairly large surface area to volume ratio due to numerous wrinkles, nooks, and crannies on their surfaces. A hand lens or a 30× handheld microscope will readily reveal the microworld detail of the surface. Students may also use their senses of touch, smell, and taste to describe the raisins. As an optional follow-

up experiment, a balance and graduated cylinder could be used to measure the raisins' average mass and volume (by water displacement) and compute their average density. Seven-Up (and other clear, colorless, sweet carbonated soft drinks) consists of water with dissolved sugar, carbon dioxide gas, and flavoring. The carbon dioxide gas is put in under pressure (i.e., is supersaturated) and when the top of the can or bottle is opened, some of the dissolved gas comes out of solution (Henry's law), which can be heard as well as seen as bubbles rising to the surface of the liquid. If a 100 mL graduated cylinder or tall, small-diameter vase is used, the students may notice that the size of the bubbles increases as they rise to the surface (due to the inverse relationship between the pressure and volume of a gas, or Boyle's law). The idea of carbonating water goes back to Joseph Priestley in the late 1700s (see Internet Connections: Wikipedia: Priestley). The point here is that there is much to be learned from carefully observing what appear to be simple systems.

2. Most people will likely miss one or more of the probable outcomes. The raisins might do one of the following:

   a. sink (if denser than the 7-Up) or float (if less dense). [*Physics:* changing buoyancy of the raisin-gas system actually causes an oscillating pattern of rising and falling]

   b. dissolve in (and/or chemically react with) the 7-Up. [*Chemistry:* the solubility of solids and gases in liquids vary at different pressures and/or temperatures]

   c. swell if water enters through the "skin" from the soft drink or shrink if water leaves the raisins. [*Biology:* osmosis predicts the former as water moves from a region of relatively higher to lower concentration, but a colored substance also diffuses out of the raisin into the soft drink]

   d. cause an increased rate of bubble formation in the 7-Up. [*Earth Science:* by providing a site of nucleation for dissolved carbon dioxide gas to come out of solution]

4 and 5. Possible variables to consider would be the number of raisins used at one time; their age (i.e., degree of dryness); the type of clear, carbonated soft drink used (versus water); and the temperature of the soft drink. When dropped into the carbonated

318

soda, the raisins first sink (because they are initially denser than water), collect gas bubbles on their irregular surfaces, then rise (because the gas bubbles increase the volume of the raisin-gas system disproportionately more than they increase the system's mass, and therefore the attached bubbles lower the density of the raisin-gas system). When the raisin–gas bubble units reach the surface, the raisins sink back down as many of the gas bubbles are released into the atmosphere. This cycle continues, but at a decreasing rate, because the concentration of available dissolved carbon dioxide decreases over time until the soft drink "goes flat." Additionally the raisins gradually swell in size and become smoother and plumper (due to diffusion of water into the raisin through its semipermeable "skin"—that is, osmosis occurs) and the 7-Up solution becomes colored as natural pigments dissolve in and diffuse out into the 7-Up solution. One possible model to construct would be small Play-Dough or clay "raisins" that can be made to be either smooth (and will sink and remain at the bottom) or dimpled (by using the head of a pin; these will rise and sink like the raisins) in a carbonated drink.

6. Further experiments for at-home play include (a) varying the temperature of the 7-Up and comparing speed and duration of cycle (gas solubility in water decreases with increasing temperature), (b) comparing the rate of bubble release with and without raisins, (c) using highly carbonated champagne-style sparkling cider versus 7-Up, (d) using Alka-Seltzer and water or baking soda and vinegar as sources of carbon dioxide, (e) comparing peeled grapes and unpeeled grapes (the surface of an unpeeled grape is better at collecting gas bubbles and will typically exhibit the oscillation pattern while lighter, peeled grapes will remain sunken), (f) using different types of dried fruit or grains that are slightly denser than soft drinks and have surfaces that could capture CO_2 bubbles (e.g., try craisins, dried blueberries, spaghetti, popcorn kernels, and Vaseline-coated raisins), (g) challenging students to keep the raisins afloat without a dissolved gas (e.g., using denser sugar or salt solutions or deforming the raisin into a boat shape), and (h) comparing the oscillation patterns of raisins that are placed in an opened bottle, an opened and re-capped bottle, and an opened bottle with fizz-saver top (in accord with

Henry's law, the fizz saver can limit the amount of dissolved gas that comes out of solution by increasing the external pressure). (*Safety Note:* Older demonstration books suggest using mothballs [either paradichlorobenzene or naphthalene], but these both have potential health risks when used in student hands-on explorations. In other words, **do not** use mothballs.)

7. Real-world applications of systems that can change their buoyancy include diving beetles (see Internet Connections: Diving Insect), fish swim bladders, hot air balloons, raising sunken ships with helium balloons, scuba diving, and submarines Also, surface area to volume effects, osmosis, and nucleation applications are ubiquitous in the biological and Earth sciences.

Activity 32

Cartesian Diver:
A Transparent but
Deceptive "Black Box"

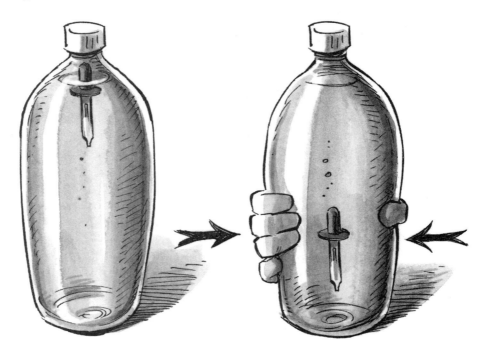

Expected Outcome

An eyedropper "diver"—*eyedropper, dropper, diver,* and *Cartesian diver* are used interchangeably in this activity—can be made to float, sink, or remain suspended in the middle of a water-filled, capped, plastic 2 L soda bottle "on command" when the bottle is secretly squeezed or released. A standard, overly simplistic, and incorrect textbook explanation is challenged.

Science Concepts

The Cartesian diver gets its name from Rene Descartes (1596–1650), the French philosopher ("I think, therefore I am") and mathematician (*x, y, z* Cartesian coordinate system of graphing). Although the diver is named after Descartes, there are other scientists whose work best explains its operation (i.e., the Greek mathematician-engineer Archimedes [287–212 BC], the French mathematician-physicist Blaise Pascal [1623–1662], and the English physicist-chemist Robert Boyle [1627–1691]). In this deceptively simple system, the diver's changing status as floating, sinking, or hovering in-between depends on several related, but distinct, scientific concepts: (1) relative density (mass/volume of the diver relative to the liquid), (2) Archimedes' principle (any object, wholly or partly immersed in a fluid, is buoyed up by a force equal to the weight of the fluid displaced by the object), (3) Boyle's law (the volume of a fixed mass of a gas is inversely proportional to the pressure acting on it), and (4) Pascal's principle (pressure exerted on a fluid in a closed vessel is transmitted undiminished throughout the fluid and acts at right angles to all the surfaces it touches).

Science Education Concepts

The Cartesian diver serves as a fairly transparent black box–nature of science system to explore how focused observations (empirical evidence) and critical questions can lead (via logical argument and skeptical review) to an explanation that is at odds with the standard (incorrect) textbook explanation. The system also serves as a *visual participatory analogy* for how science instruction can catalyze cognitive construction by recognizing that different learners may react differently to the same "pressure" and may need differentiated environments to learn. It also shows how science can be considered a form of planned, purposeful playfulness.

Materials

Each group of two to four learners will need the following:

- 1,1 L or 2 L empty plastic soda bottle

- 2 eyedroppers (either the standard glass-and-black-rubber-topped or the all-plastic, completely transparent variety; the all-plastic eyedroppers require brass nuts as weights to stabilize them) (*Note:* The standard, black-rubber-topped droppers are better for the teacher "magic" demonstration [see Procedure]. The all-plastic droppers allow learners to see the change in volume of trapped air and are better for the hands-on explorations.)
- Plastic drinking cup. The cups should be taller than the length of the droppers.

Points to Ponder

Everything should be made as simple as possible... but not simpler.... A theory is more impressive the greater is the simplicity of its premises....

—Albert Einstein, German-American physicist (1879–1955)

Reflective thinking is always more or less troublesome because it involves overcoming inertia that inclines one to accept suggestions at their face value; it involves willingness to endure a condition of mental unrest and disturbance.

—John Dewey, American philosopher-educator (1859–1952) in *How We Think* (1910)

Procedure

(See answers to questions in steps #1c, #2, #4a, and #4b on p. 329.)

1. If desired, you can begin this activity as a science "magic trick" demonstration.

   a. Before the presentation (and out of view), completely fill a 2 L bottle with room temperature water (slightly colored with food coloring if desired for contrast with the water in the eyedropper), insert an eyedropper that contains a ratio of air/uncolored water that allows it to just barely float in an upright

position, and cap the bottle. It's easiest to find the proper ratio by varying the amount of water in the eyedropper and testing how it floats in a large plastic cup from which it can be easily retrieved before inserting it into the bottle.

b. Announce to the class that you can "order" the eyedropper to dive on your command. Hold the bottom of the bottle with your left hand and apply pressure by surreptitiously squeezing from behind as you pass your right hand (as a distraction) along the right side of the bottle without touching the bottle. Announce to the eyedropper, "I think you will sink now." Depending on how much pressure you apply, you can get the eyedropper to continue to sink or remain suspended at any depth. When you release pressure, it will float back to the top.

c. Ask: Is this magic or science? How can we tell the difference? What science concepts can account for this behavior? Compile possible answers, but do not discuss them until after steps #2–#4, below.

2. Allow time for hands-on exploration of this discrepant-event system. Give each group a tall plastic tumbler of water and two identical small eyedroppers, and ask the group members to make one eyedropper float with its top barely above the surface of the water and the other eyedropper sink. When the students achieve this task, ask them to carefully observe what is different between the two systems.

3. Have the groups transfer the two eyedroppers to a 1 or 2 L soda bottle filled with room temperature water without changing the volumes of trapped air and water. Have them put the caps tightly on their bottles and see if they can repeat your "trick" with the floating/sinking dropper.

4. As groups figure out that they need to apply external pressure to the sealed bottle, ask them to compare the two eyedroppers when the floater has sunk versus when it is still floating. Ask them to consider the following questions:

a. Is the Cartesian diver an open or closed system with respect to the water and to the air?

b. How is it like and unlike a submarine? Is the key variable the volume of trapped air or the mass of water in the eyedropper? What is constant and what changes in these two systems?

Debriefing

When Working With Teachers

Use the Einstein quote and the fact that this "simple" activity is commonly explained incorrectly to point to the importance of teachers not simply accepting all textbook explanations. An analogy can also be made that student performance is related to the degree of "pressure" and "support" provided by the teacher via his or her integrated curriculum-instruction-assessment system. Differentiation is needed because different students will respond differently in the same environments (e.g., note the behaviors of different divers in various liquids or in the two different-shaped bottles in Extensions #1 and #2). The key is to use an intentionally designed, scaffolded sequence of activities (e.g., the 5E Teaching Cycle: Engage, Explore, Explain, Elaborate, Evaluate; see Appendix B for an explanation) to catalyze student construction (or reconstruction) of a network of related science concepts. It's also useful to point out the multiple benefits of teaching with simple toys (e.g., motivation, low cost, and real-world relevance). Interested teachers can explore the research on student misconceptions related to gases and/or density–floating/sinking (e.g., see Driver et al. 1994; chapters 9 and 13 and/or chapters 8 and 12 are good starting points).

When Working With Students

The "magic trick" part of this activity makes for a nice Engage-phase teacher demonstration, to be followed by the hands-on Explore phase. During the Explain phase, focus on getting students to clearly define the "system" and noting the mathematics of density $(D = M/V)$ when you have a fixed mass and variable volume of air. Many variations (e.g., temperature effects) and real-world applications could be examined in the Elaborate and Evaluate phases. Given that Boyle's law, the idea of open and closed systems, and the mathematics of fractions are all key to the explanation, a Cartesian diver is probably best explored at the grades 5–12 level—despite the fact that it makes for such a fun, hands-on science toy at much earlier grade levels.

Extensions

1. *"Lighter" Liquids, Less Support.* The teacher can ask the students to predict-observe-explain how the behavior of a water-and-air-filled diver that just floats in a water bottle might change if it were placed in a bottle of lower density liquid such as oil or alcohol. The teacher can demonstrate that the amount of air would need to be increased (and water decreased) in the diver to get it to just barely float inside bottles filled with either of these less-dense liquids. Follow this demonstration by asking the students to predict-perform-explain what they need to do to get a water-and-air-filled diver to work in a more-dense fluid such as concentrated saltwater or sugar water. They will discover that they need to decrease the amount of air (by increasing the amount of water) to get the diver to just float in a vertical position in a denser liquid (and quickly sink with external pressure). They can also consider how fluid temperature would affect the density of the surrounding liquid (i.e., density decreases with increasing temperature as the molecules move further apart and the volume of the fluid increases and the mass remains constant).

2. *Discrepant Detergent Bottle.* If a thin flat bottle is used (e.g., a dish detergent or window cleanser bottle—or an oval cross section of that type of bottle), it is possible to make a sunken, just-a-little-too-dense diver float by applying pressure on the two narrow sides and distorting the bottle into a more cylindrical, larger-volume shape. This decreases pressure on the diver and allows the sealed air mass to expand, increasing the volume of trapped air in the diver and decreasing the diver's density, allowing it to float as long as the bottle stays in this distorted shape.

3. *Soy Sauce Submarines.* Translucent packets of soy sauce (with a trapped air bubble) and other small sealed packets of condiments can be used in lieu of an eyedropper. Find a packet that just barely floats and place it in the Cartesian bottle. When the bottle is squeezed, the packet will sink. Students can observe the change in air volume without the added variable of water seemingly being added to the relevant system. This extension can be used to help students understand that the volume of air is the key variable.

4. *Magnetic Movements Matter.* For a unit on magnetism, several iron nails can be inserted into a cork so that it just barely floats (or if a little heavier, just sinks) in a 2 L soda bottle that has its top cut off (to form a plastic beaker). If a strong magnet is placed in your hand and covered with a glove, you can make the cork-and-nails "octopus" move on command when you move your gloved hand along the sides of the bottle.

5. *Density Discrepancy Toys.* Science education supply companies sell a variety of products for simple and enhanced products related to Cartesian divers. Educational Innovations (*www.teachersource. com.* 888-912-7474) sells the following products:

   - Graduated Transfer Pipettes: #PP-222A. $10.95/100 plastic pipettes used in conjunction with Cartesian Diver Ballast Nuts: #CD-3. $16.50/100 brass or stainless steel nuts
   - Catch and Release Cartesian Diver: CDH-413: $4.50
   - Fizz-Keeper Pump (to apply pressure to a 1 L or 2 L bottle): #CD-4: $3.50
   - See also related book: *39 Fantastic Experiments With the Fizz Keeper* by Brian Rohrig, BK-350. $17.95

   Steve Spangler Science (*www.stevespanglerscience.com/product/ 1164*) sells products similar to those above.

6. *Real-World Relevance.* Students can investigate how fish (with air bladders), diving beetles, human divers, and submarines use the ideas in this activity to control their depth (see Internet Connections: HowStuffWorks). Hot air and lighter-than-air balloons that float in the "sea of air" of our atmosphere are based on the same scientific principles. See Activity #14, Extension #1 in this book.

Internet Connections

- Arbor Scientific's Cool Stuff Newsletter: *www.arborsci.com/CoolStuff/Archives3.aspx* (See Chemistry: Gas laws and Pressure & fluids for demonstrations, including the Cartesian diver.)

- Becker (chemistry) Demonstrations and experiments (with Quick-Time movies): (1) Cartesian diversions: Basics: *http://chemmovies.unl.edu/chemistry/beckerdemos/BD031.html* (2) Closed system divers: *http://chemmovies.unl.edu/chemistry/beckerdemos/BD034.html* (3) Counters and messages: *http://chemmovies.unl.edu/chemistry/beckerdemos/BD032.html* (4) Diving whirligigs: *http://chemmovies.unl.edu/chemistry/beckerdemos/BD033.html* (5) Retrievers: *http://chemmovies.unl.edu/chemistry/beckerdemos/BD035.html*

- Cartesian Diver Java Applet (allows user to control external pressure and diver location): *www.lon-capa.org/~mmp/applist/f/f.htm*

- Diving insect regulates buoyancy: *www.adelaide.edu.au/adelaidean/issues/12121/news12122.html*

- Exploratorium Snackbook: *www.exploratorium.edu/snacks/condiment_diver/index.html* and *www.exploratorium.edu/snacks/descartes_diver/index.html*

- HyperPhysics, Department of Physics and Astronomy, Georgia State University: Videos/demos: Cartesian diver: *http://hyperphysics.phy-astr.gsu.edu/hbase/pbuoy3m.htm/#c1*

- HowStuffWorks: Submarines: *http://science.howstuffworks.com/submarine1.htm*

- A Philosophical Toy: A teacher explores the history and "alternative" explanations: *www.ed.uiuc.edu/courses/CI241-science-Sp95/resources/philoToy/philoToy.html*

- Purdue University, Division of Chemical Education: Cartesian diver (with QuickTime movie): *http://chemed.chem.purdue.edu/demos/main_pages/1.2.html*

- University of Iowa Physics and Astronomy Lecture Demonstrations (video clips): Heat and fluids: *http://faraday.physics.uiowa.edu* (five demonstrations on density and buoyancy)

- Wake Forest University Physics Department: Physics of matter videos: Cartesian diver: *www.wfu.edu/physics/demolabs/demos/avimov/bychptr/chptr4_matter.htm*

- Whelmer: #49: Potato float (layer two liquids): *www.mcrel.org/ whelmers/whelm49.asp* and #64: Liquid rainbow (layer five liquids by density): *www.mcrel.org/whelmers/whelm64.asp*
- Wikipedia: (1) Boyle's law (and animation): *http://en.wikipedia. org/wiki/Boyle%27s_law* (2) Buoyancy and Archimedes' principle: *http://en.wikipedia.org/wiki/Archimedes_Principle* (3) Pascal's principle: *http://en.wikipedia.org/wiki/Pascal%27s_principle*

Answers to Questions in Procedure, steps #1c, #2, #4a, and #4b

1c. Science explains natural phenomena in terms of empirical evidence (i.e., measurable quantities of matter and energy), logical argument, and skeptical review. In this case, the movement of a mass (the dropper) implies a force of some kind and, therefore, the expenditure of energy or work. All of the Science Concepts discussed at the start of this activity are relevant to explaining the floating, sinking, and hovering behavior of the diver under variable amounts of external pressure.

2. The floating dropper has a smaller volume of water and larger volume of air relative to the dropper when it is sunken.

4a. The diver is an open system with respect to water movement and closed with respect to air.

4b. Both a submarine and a Cartesian diver can "take on" additional water, but the Cartesian diver (eyedropper) is never sealed off from the surrounding water as a submarine can be. The additional water added to ballast tanks adds to the submarine's mass and overall density (its overall volume is a constant). When pressure is applied to the bottle, water is forced into the eyedropper. It compresses the trapped air and reduces the volume of the constant mass system of the eyedropper and its fixed mass of air. Because the system is open with respect to water, the mass of the water "inside" the dropper is not relevant. As the volume of the diver system is reduced, its density increases and it sinks ($D = M/V$). (*Note:* Many science demonstration books and articles fail to note this distinction and focus attention (incorrectly) on the mass of water rather than the changing volume of a fixed mass of trapped air.)

Crystal Heat: Catalyzing Cognitive Construction

Expected Outcome

A small metal clicker/disk within a sealed pouch containing a clear, colorless liquid is clicked, and the liquid rapidly crystallizes into a solid mass and releases a large quantity of heat. If the solidified pouch is allowed to cool back to room temperature and then is reheated, it returns to its liquid state with its "recharged" latent heat available for release again when desired.

Science Concepts

Some crystalline solids are hygroscopic (they absorb and weakly bond to water molecules) and have the ability to form supersaturated solutions in their own water of hydration. In this activity, sodium acetate trihydrate melts (dissolves in its own water of hydration) when heated, and it remains stable as a supersaturated liquid solution at room temperature until it is disturbed by either clicking the metal disk (if done in a sealed pouch) or adding a seed crystal (if done in a test tube). Either the metal clicker or the seed crystal can be viewed loosely as a "catalyst" for accelerating the rate of the physical change.

(*Note:* Technically, a catalyst is substance that facilitates a particular chemical reaction by lowering the activation energy barrier by providing an "active site" where bonds of the "old" reactants can be more readily broken and bonds of the "new" products can form. Catalysts speed up the rate of chemical reactions that are thermodynamically favored (but kinetically slow) without being used up or permanently changed in the process. Typically, catalysts lower the activation energy for a given reaction, but inhibitory catalysts also exist, especially in biochemical systems.)

The heat needed to melt and dissolve the crystals in their own water of hydration is stored as latent heat (a type of chemical potential energy) that is released when the solution recrystallizes. This repeatable cycle qualitatively demonstrates the law of conservation of energy (solid + heat ←→ liquid; energy is similarly conserved in any liquid + heat ←→ gas phase change). The phase change in this activity is represented by the equilibrium reaction:

$$NaC_2H_3O_2 \cdot 3H_2O \text{ (solid phase)} + 4 \text{ kilocalories/mole} \longleftrightarrow Na^+ + C_2H_3O_2^- + 3H_2O \text{ (liquid phase)}$$

On Earth, the fact that water exists and reversibly changes between all three phases is central to the interactions between the lithosphere, hydrosphere, atmosphere, and biosphere. For example, water's liquid-to-gas phase change accounts for the cooling effect of animal perspiration (see Extensions, step #4). The reverse change of gaseous water condensing accounts for the tremendous release of energy via thunder and lightning during rainstorms as well as the terrible burns that can be caused by steam. (*Note:* The sealed pouch can also be used to demonstrate the conservation of matter during a physical change of state or phase—that is, it has the same mass before and after the change.)

Science Education Concepts

Reusable heat pouches can be used as a *visual participatory analogy* that models how learners respond to integrated curriculum-instruction-assessment plans that help them catalyze and crystallize cognitive constructions. (See Answers to Questions in Procedure, step #5, pp. 341–342, for an extended discussion of this analogy.)

The sealed pouches also serve as an economical, portable, re-usable system that demonstrates two themes of the AAAS Benchmarks: systems and constancy and change (AAAS 1993). Additionally, the activity models the use of a chemically safe (i.e., sodium acetate is the primary flavoring in salt and vinegar potato chips), environmentally friendly, reusable, one-time investment that has a practical real-world application.

Materials

- 2 heat pouches (at a minimum) for demonstration purposes; 12 pouches for hands-on explorations by a class. Available from science suppliers such as
  - Arbor Scientific. *www.arborsci.com/detail.aspx?ID=426* 800-367-6695. Cristal Heat pouch (with an accompanying lab manual): 4 in. × 4 in. P3-1015. $10.00 (or $8 without the manual)
  - Educational Innovations. *www.teachersource.com* 888-912-7474:
  - Chemical Heat Pack: 4 in. × 4 in. HEA-400. $6.95 each
- Ice cubes (Procedure, step #1)

An essentially identical product—an instant reusable heat pouch—is often sold in drugstores or camping stores. (Do not confuse these heat pouches with reusable, microwavable gel packs, which do not undergo a phase change.)

Alternatively, sodium acetate trihydrate crystals (melting point: 58°C or 136°F) are available from any science or chemical supply company. These crystals can be heated and melted in a *clean, dry*

Safety Notes

1. Students and teacher should wear indirectly vented chemical splash goggles during this activity.

2. Sodium trihydrate pentahydrate and sodium acetate are hazardous chemicals (toxic if swallowed and can cause skin and respiratory irritation). Review safety information from MSDSs with students.

flask or test tube, lightly stoppered (until the liquid returns to room temperature), and then firmly stoppered (until a seed crystal is added). (See Internet Connections: Doing Chemistry and Purdue University.) These sites provide more detailed directions and video clips on how to do this variation on the demonstration. Sodium thiosulfate pentahydrate or photographer's hypo (mp 48°C) is another common crystal that exhibits supersaturation. In either case, the first step of the Procedure makes use of ice cubes pulled from a container of ice water.

Points to Ponder

Most modern chemistry curriculums attempt to do too much too soon…. Sampling the fountain of knowledge from a modern chemistry course is rather like trying to get a drink from a fire hydrant.

—Henry Bent, Jr., chemist and chemistry educator, in the *Journal of Chemical Education* (January 1980)

Chemistry is a foreign language twice over, strange terms for strange things.

—Henry Bent, Jr., chemist and chemistry educator, in the *Journal of Chemical Education* (September 1984)

Procedure

(See answers to questions in steps #1–#3 and #5 on pp. 341–342.)

1. Introduce the activity by asking, What substance is both the most common chemical on Earth and one of least common chemicals in our solar system? What are some of the unique properties of water compared to other common chemicals on Earth?

    Take an ice cube from a container of ice water and ask: What is needed to turn the solid ice cube into liquid water? As the heat from your clenched hand melts the ice, ask: If heat energy is needed to melt a solid and return it to the liquid state, then what must happen to some of the internal energy in liquid water for it

334

to turn back into a solid during freezing? How do you think the amount of energy required to melt ice compares to the amount of energy liquid water needs to "lose" to become solid ice?

2. Hold up a liquefied Cristal Heat pouch or other heat pouch and tell the learners that this system will be used to study the idea of phase changes and "internal energy." Ask one or more volunteers to confirm that the liquid-containing pouch feels as if it is at room temperature. Ask the learners: What might we do to crystallize this clear, colorless liquid? After someone offers that it can be "cooled down" by putting it in a freezer, suggest that there is another, quicker way. Distribute several of the pouches and ask the learners to examine and play with the metal clicker/disk (without opening the pouch!). As soon as they click the disk, the liquid will rapidly begin to crystallize, releasing a considerable amount of heat. When this happens, it will come as quite a visual and tactile surprise! Be sure that the pouches keep circulating around the room so everyone gets a chance to see and feel the result of the phase change.

3. Write a simplified summary of the physical change: liquid → solid + heat. Suggest that the action of the clicker/disk can be viewed loosely as a "catalyst" to initiate the physical change of phase or state of matter. Ask: Do you think this process is reversible (as it is with solid ice and liquid water)? What might we do to return the now solid pouch to its previous liquid state?

   Heat the solid pouch in boiling water until all of the solid turns into liquid. Then, let the liquid pouch cool to room temperature before repeating the clicking-to-solidify process. If you have multiple pouches, you can alternate quickly between the solid and liquid states. Ask: Which phase of matter contains more internal energy, the solid or the liquid state? How does this phase change (like that of ice + heat ←→ liquid water) provide qualitative evidence for the law of conservation of energy?

4. *Optional*: If time permits, also perform the liquid → solid phase change by adding a "seed" crystal of sodium acetate trihydrate to a Pyrex flask, test tube, or petri dish that contains the room temperature supersaturated liquid. The latter can be displayed on an overhead projector or with a document camera so that the crystallization process is visible to everyone. Alternatively,

a crystal "sculpture" can be made by using a laboratory "water" bottle containing the liquid sodium acetate to add to a small mass of the solid. For details, see Internet Connections: Doing Chemistry and Purdue University. (*Note:* Be sure to keep applying pressure to the bottle and moving it to prevent crystals from forming inside the bottle.)

Questions When Working With Teachers

5. Ask: How might this clicker-catalyzed, reversible phase change system serve as a *visual participatory analogy* for the interrelated processes of teaching and learning? How are the processes of learning unlike the processes of teaching?

Debriefing

When Working With Teachers

Discuss one or more of the Science Education Concepts on page 332. Note that our conceptions are continually, as it were, being dissolved or melted and recrystallized in light of our new experiences or stresses to our preexisting mental schemata. Use the Henry Bent, Jr., quotes to emphasize the need to cover a smaller number of central topics or "big ideas" better. Urge science teachers to make the familiar world of everyday macroscopic events—such as the phase changes of water—seem "strange" (i.e., worth noticing and exploring) and to make the strange, unseen world—such as molecular interactions—seem "familiar" (or "sensible" in both meanings of the word). Computer simulations (see Internet Connections: PhET) are especially useful for helping learners to visualize the otherwise invisible world of the kinetic molecular theory.

Be sure to note that the crystallization process is a physical change and that the metal clicker (or seed crystal) technically is not a catalyst. However, an analogy can be drawn between the crystallization process and the teaching-learning dynamic. Learning is a natural and continuous biochemical, neurological, and psychological process of change that can be greatly accelerated in the presence of an effective "teacher-catalyst." Although teachers are not brain surgeons, they do support the minds-on reconstructions (or "sculpting" or "rewiring") of learners' neurological networks. It should also be noted that the best teachers are in fact changed *themselves* as a result of their interactions with learners (but never, one hopes, "used up" or "burned out"). See Answers to Questions in Procedure, step #5, pp. 342, for a more extended discussion of this analogy.

When Working With Students

The activity can be used at the middle school level to introduce the idea of phase change and conservation of energy. If the activity is used during the Engage phase, do it more as a science magic trick, followed by Explore-phase, hands-on experiments (e.g., heating and cooling curves). (For a discussion of the five phases of one particular teaching cycle, see Appendix B.) At the high school chemistry level, the demonstration could also lead to more quantitative Explore-phase experiments (e.g., measure the heat evolved per pouch, determine the mass of the supersaturated solution inside the pouch, empirically determine the latent heat of crystallization for sodium acetate) and to more detailed molecular explanations. In either case, avoid prematurely introducing unnecessary science terminology until the Explain phase, during which molecular animations can be used to help learners see "what's really going on." (*Note:* The relative distances between molecules in the gaseous phase are, by necessity, always misrepresented in both animations and static images.)

In the Elaboration phase, any number of real-world applications can be considered that relate specifically to technologies involving physics and chemistry (e.g., refrigeration and air conditioning units that use the liquid + heat ← → gas phase change to "pump" heat "uphill" from a cooler to a warmer place) or biology and Earth science–related phenomena (see Extensions). At the middle school and high school levels, the commercial product serves as an example of a system that absorbs or releases heat energy in a reversible physical change where both matter and energy are conserved.

Extensions

1. *Dry Ice Is Nice, but It's Only a Phase.* Dry ice, or solid carbon dioxide, displays the property of *sublimation* as it changes from the solid state directly to the gaseous state without first becoming a "wet" liquid. This property can be easily demonstrated by placing a solid piece of dry ice into a deflated balloon, sealing the

balloon, and watching it "self-inflate." Again, the solid absorbs heat from the environment (an endothermic process) to make the phase change to the gaseous state (see Internet Connections: Doing Chemistry). Similarly, iodine crystals placed in a stoppered flask sublime (accelerated by gentle warming) to form a denser-than-air, purple-colored gas without going through the typical, intermediate liquid phase. Mothballs also sublime.

2. *Changes, Catalysts, and Combustion.* A sugar cube will brown and melt (sucrose or $C_{12}H_{22}O_{11}$ melts at 185°C) but not catch fire when the flame from a Bunsen burner or propane torch is held on it. If the sugar cube is rubbed in fine-powdered, activated charcoal (carbon) or wood ash, it will readily ignite and continue to burn. Each cell in the human body "burns" glucose [$C_6H_{12}O_6 + 6\,O_2 \rightarrow 6\,CO_2 + 6\,H_2O$ + energy] in a catalytically controlled series of chemical reactions (called cellular respiration) that happen at body temperature (37°C) without a flame! Most biochemical reactions are controlled by special proteins called enzymes that act as catalysts. Truly, we should CELLebrate catalysts!

3. *Dazzling Demonstrations.* Videos and written instructions at the Instructables and Woodrow Wilson websites (see Internet Connections) depict two different and equally dramatic versions of the same demonstration on the catalyzed decomposition of 30% hydrogen peroxide ($2\,H_2O_2 \rightarrow 2\,H_2O + O_2$ + energy). Like the use of ash in the burning of a sugar cube, the manganese dioxide used in this chemical reaction is a true catalyst.

A commercial example of a catalyst-accelerated chemical reaction is the drugstore-variety hand warmers. The most common version relies on the rapid oxidation of iron to release heat. These hand warmers can be compared to the much slower, exothermic rusting of a piece of steel wool that has been dipped in vinegar and placed in a Styrofoam cup with a lid and thermometer. (*Note:* Larger, one-time-use heat pouches typically have two separate pouches that contain calcium chloride and water, which, when mixed, create an exothermic physical change [dissolution]. Heat-related experiments with these systems can be performed at a much lower cost by buying calcium chloride in bulk in the form of low-temperature-acting, winter de-icing road salt. Check the label on the 5–10 lb. bags to make sure you are

Safety Notes

1. Students and teacher should wear indirectly vented chemical splash goggles during this activity.

2. Do Extension #2 under a fume hood.

3. See Extension #3:

- 30% hydrogen peroxide is a strong oxidizing agent. Contact with eyes should be avoided. In case of contact, flush with water for at least 15 minutes. Get medical attention if eyes are affected. Also avoid contact between hydrogen peroxide and combustible materials. 30% hydrogen peroxide must be stored in its original container.

(cont.)

buying *calcium chloride*, not the cheaper *sodium chloride*, which doesn't cause release of heat when it is dissolved in water.)

4. *Perspiration, Thermoregulation, and Sweating.* Water has a very high heat of vaporization and condensation (540 calories/gram). When animals perspire (and/or pant), heat leaves their bodies as liquid water and is converted into water vapor. If the humidity is too high, perspiration remains on the skin in the liquid phase (as sweat) where it cannot play its normal cooling function. Alcohol rubs also have a cooling effect because alcohol evaporates even faster than water. Learners can test this effect by using a cotton ball to rub room temperature rubbing alcohol on the back of their hands or by placing the same cotton ball on a laboratory thermometer. Pigs wallow in the mud for both its cooling and sun-protection effects. The high heat of condensation of water (the "heat of vaporization") also explains why steam burns have such a severe effect on exposed skin: The steam returns to the liquid state by releasing the heat of condensation.

5. *Supersaturated Is Sweet.* Honey is an example of a naturally occurring supersaturated solution that can solidify if it sits on a shelf for a long time or if "seed" crystals are introduced into the jar. Learners can investigate the honeybee industry and learn about how honey is naturally produced and commercially harvested. Biology classes may wish to investigate the recent colony-collapse disorder that has resulted in significant declines in the number of pollinating bees. These bees are responsible for an estimated 15–30% of the food U.S. consumers eat, including apples, beets, and cabbage. See Internet Connections: Wikipedia: List of Crop Plants Pollinated by Bees, which includes an extensive and impressive table.

Safety Notes

3. See Extension #3 (cont.):

- Manganese dioxide is an extremely hazardous chemical. If desired, other catalysts can be used for this reaction—for example, potassium iodide (KI), active dry yeast, and even raw liver! **Like all chemistry demonstrations, this one requires practice and careful attention to safety guidelines by a knowledgeable teacher.** If the necessary knowledge, reagents, and/or proper equipment are lacking, dramatic online videos are preferred to live, uncontrolled, potentially hazardous reactions.

4. In Extension #4, use very small amounts of rubbing alcohol in a well-ventilated area. Alcohol is extremely flammable. Keep away from heat source or flame.

Internet Connections

- Bizarre Stuff You Can Make in Your Kitchen: *http://bizarrelabs. com/index.htm* (Go to Sugar cube tricks: The burning sugar cube.)

- Concord Consortium (free downloadable simulations): Phase change: Molecular Workbench Software Homepage: *http:// mw.concord.org/modeler/molecular.html*

- Doing Chemistry: Movies of chemistry demonstrations: Super-saturation: *http://chemmovies.unl.edu/Chemistry/DoChem/DoChem058.html* and Dry ice in a balloon: *http://chemmovies.unl.edu/chemistry/dochem/DoChem087.html*

- Instructables: Genie in the bottle demonstration video (decomposition of hydrogen peroxide): *www.instructables.com/id/Genie-In-A-Bottle*

- PhET Interactive Simulations: States of matter: Molecular view with temperature, pressure, and volume controls and phase changes for three phases of Ne, Ar, O_2 and H_2O: *http://phet.colorado.edu/simulations/sims.php?sim=States_of_Matter*

- Purdue University: Sodium acetate liquid → solid (supersaturation and crystallization demonstration): *http://chemed.chem.purdue.edu/demos/main_pages/15.2.html*

- University of Minnesota, Chemistry Department: Genie in a bottle: *www.chem.umn.edu/services/lecturedemo/info/genie.htm*

- Wikipedia: (1) Catalysis: *http://en.wikipedia.org/wiki/Catalysis* (2) Hand warmers (both types): *http://en.wikipedia.org/wiki/Hand_warmer* (3) Honey: *http://en.wikipedia.org/wiki/Honey* (4) Perspiration: *http://en.wikipedia.org/wiki/Perspiration* (5) Pollinator decline: *http://en.wikipedia.org/wiki/Pollinator_decline* (6) Sodium acetate: *http://en.wikipedia.org/wiki/Sodium_acetate* (7) Sodium thiosulfate: *http://en.wikipedia.org/wiki/Sodium_thiosulfate* (8) Supersaturation: *http://en.wikipedia.org/wiki/Supersaturated* (9) *http//en.wikipedia.org/wiki/List_of_crop_plants_pollinated_by_bees*

- Woodrow Wilson Leadership Program in Chemistry: Demonstration of the rapid MnO_2 catalyzed decomposition of 6% H_2O_2 in the presence of liquid detergent (for foaming): *www.woodrow.org/teachers/chemistry/institutes/1986/exp21.html*

Answers to Questions in Procedure, steps #1–#3 and #5

1. Water is both the most common chemical on Earth and one of the least common substances (along with biomolecules like DNA) in our solar system. From a chemical perspective, its numerous "uncommon" properties (given that it has a low molecular weight) are related to its high degree of intermolecular hydrogen bonding. These unique properties include having a high specific heat, heat of vaporization, boiling point, and surface tension and—the most uncommon trait—a solid phase that floats (rather than sinks) in its own liquid.

   What is relevant for this activity is the "simple," everyday observation that water exists in all three readily convertible phases under normal Earth-environment temperatures and pressures. In nature, when liquid water in lakes and streams freezes, it releases heat to the environment. Freezing is an *exothermic* process; melting is an *endothermic* process. The amount of heat energy exchanged in these reversible processes is identical (as predicted by the law of conservation of energy).

2. Clicking the disk initiates the phase change from a liquid to a solid. The chemical heat pouch serves as a small-scale model of the reversible process of solid + heat ←→ liquid phase change.

3. The liquid phase of any given substance contains more internal energy than the solid phase of the same substance (even if the substance is at the same temperature in both phases). It is this latent heat that is released when the liquid crystallizes or freezes. The heat added to melt a crystalline solid is equivalent to the heat removed or released when a liquid crystallizes—again, in keeping with the law of conservation of energy. A "big idea" in Earth science is that the existence of water in all three phases on Earth provides a thermo-regulatory mechanism that helps moderate the Earth's climate (i.e., it provides a kind of homeostatic thermostat for both local environments and the planet as a whole). Water is not just a key chemical found in all living cells; it is the key chemical for regulating the entire Earth's biosphere!

4. [There are no questions or answers for step #4.]

5. *Visual Participatory Analogy for Teaching-Learning:* Especially if teachers are familiar with the chemistry behind how catalysts

work, this is a conceptually rich analogy. The liquid pouch can be used to represent learners' preinstructional, less organized state of understanding. The teacher's curriculum-instruction-assessment plans and efforts (or more simply, the "teacher") can be considered to be a catalyst (or clicker) that helps learners construct a more "crystallized" network (or schema) of interrelated concepts that constitutes a "solid" understanding.

All chemical reactions involve the endothermic breaking of old bonds and the exothermic forming of new ones. Similarly, learning is "thermodynamically favored" (i.e., it can occur without a teacher), but may be "kinetically slow" in the absence of a teacher-catalyst. Just as all life depends on cells performing untold numbers of complex, simultaneous, enzyme-catalyzed chemical reactions 24/7, similarly, conceptual change and cognitive development are lifelong processes of both unlearning (breaking old bonds) and new learning (forming new ones) as our prior conceptions are continually challenged by new experiences.

Some new experiences are readily assimilated into prior conceptual networks (or schemata), much like an extra drop of liquid sodium acetate would crystallize onto the surface of solid sodium acetate without otherwise changing the physical state of the solid mass. However, other experiences are anomalies that don't fit into our prior mental schemata. These experiences require accommodation—previous conceptualizations need to be "melted" and recrystallized in new configurations. Much like the invisible, too-small-to-be-seen changes in the spatial arrangement and energy of molecules or ionic crystals during phase changes and chemical reactions, the neurological wiring and rewiring associated with learning is not visible at the macroscopic level. All assessment efforts are indirect, inferential means of checking internal states of understanding by eliciting external behaviors tied to specific, representative tasks (i.e., assessments attempt to make visible the invisible processes and products of thinking).

Limitations of this extended analogy include the following: (a) it misrepresents the relative roles and efforts of the teacher and learner, (b) it grossly reduces and underestimates the complexity of the actual teaching-learning process, and (c) energy is required from both parties during all phases of the process. One hopes, however, that both teachers and learners are mutually re-energized in their exothermic partnership!

Appendix A

Selection Criteria for Discrepant Events and Analogical Activities

Includes Connections to National Science Education Standards

With limited time for curricular preparation and professional development, overworked teachers make strategic cost/benefit assessments before investing time and effort in learning a new teaching strategy or student activity. To help ensure that the investments teachers put into this book pay rich dividends, the activities are designed to meet six research-informed criteria established by the author. The activities are Safe, Simple, Economical, Enjoyable, Effective, and Relevant (S_2EE_2R) for both teachers and grades 5–12 students (O'Brien 1991).

1. Safe

Physical safety and psychological safety are essential conditions for effective science teaching—"Ensure a safe work environment" (NRC 1996, p. 43) (from the National Science Education Standards Teaching Standard D). What is regarded as *physically safe* depends on the environment (e.g., a regular classroom versus a laboratory with ventilation, fume hoods, and safety shields) and the knowledge, skills, and preparation of the teacher and students, as well as the actual materials and techniques employed (Flinn Scientific; Kwan and Texley 2002; Kwan and Texley 2003; Texley, Kwan, and Summers 2004). All of this book's activities can be performed as *teacher* demonstrations in regular classrooms with standard safety facilities and equipment. When used as *student* hands-on explorations, it is essential that the teacher model—and monitor—proper safety procedures and equipment both for safety purposes and to obtain useful, reliable results. In both cases, professional prudence and prior preparation are essential before doing any science activity.

Psychological safety is present in a classroom where mutual trust, respect, and support exist. In such a classroom, "miss-takes" are viewed as necessary and productive steps toward building competence and cultural and cognitive diversity are seen as assets. Teaching Standard E in the National Science Education Standards (NSES) elaborates on this point:

> Teachers of science develop communities of science learners that reflect the intellectual rigor of scientific inquiry and the attitudes and social values conducive to science learning. In doing this, teachers:
>
> - Display and demand respect for the diverse ideas, skills, and experiences of all students:
> - Require students to take responsibility for the learning of all members of the community.
> - Nurture collaboration....
> - Structure and facilitate ongoing formal and informal discussion based on a shared understanding of rules of scientific discourse.
> - Model and emphasize the skills, attitudes, and values of scientific inquiry. (NRC 1996, pp. 45–46)

Psychological safety is especially important when teachers use counterintuitive, discrepant-event activities to activate and challenge students' prior conceptions. If "getting it wrong" is ridiculed by their teachers or peers, students will not risk making public predictions or be able to make careful observations and explanations (POE) of their reasoning. When a student (of any age) fears failure or otherwise feels psychologically threatened, he or she "cognitively downshifts" (Hart 1983). As the human body's physiology shifts into the fight, flight, or freeze mode, the higher-order cognitive functions associated with learning, memory, problem solving, and creative thinking are restricted. If such stress and anxiety are chronic, a student may develop a sense of learned helplessness (Jensen 1998). Conversely, when a student's teacher and peers respond with empathy and support, the student's ability to deal effectively with potentially stressful situations and avoid "emotional hijacking" of higher cognitive functioning is enhanced (Goleman 1995, 2006).

This book is designed to help teachers directly experience that "some of what we know isn't so" (Gilovich 1991). That is especially true with respect to "unquestioned answers" related to beliefs about science, teaching, and learning. When teachers themselves—in, for example, professional development settings—experience the cognitive dissonance that their students often confront, they are in a better position to help students with such challenges.

2. Simple

NSES Program Standard B argues for "developmentally appropriate," inquiry-based science where students "learn science in a way that reflects how science actually works" rather than being expected to "learn terms and perform activities that are far beyond their cognitive and physical development level" (NRC 1996, p. 214). Teachers are challenged to facilitate students' discoveries and understanding about unusual or strange science concepts (e.g., atoms and molecules, cells, evolution, geological time, plate tectonics, and energy transformations) through interactive, experiential-learning activities that involve simple, everyday—yet often under-explored—phenomena.

The AAAS Benchmarks (AAAS 1993), the AAAS *Atlas of Science Literacy,* volumes 1 and 2 (AAAS 2001, 2007), and the National

Science Education Standards (NRC 1996) map out developmentally appropriate, conceptually scaffolded targets for K–12 students. Unfortunately, many commercial textbooks introduce sophisticated theoretical constructs prematurely, without attending to developmental appropriateness or providing adequate experiential evidence (Michaels, Shouse, and Schweingruber 2008; NRC 2007; Singer, Hilton, and Schweingruber 2006). Although hands-on explorations are an essential component of effective science programs (and over one-third of the book's activities include hands-on explorations), they are not the only component (NRC 1996; Singer, Hilton, and Schweingruber 2006). In the hands of skilled science teachers, simple, interactive, discrepant-event demonstrations and experiments can be effectively used across a range of grade levels by varying students' physical involvement, the extent of Socratic probing and cognitive scaffolding, and the intended learning outcomes.

It should be noted that simplicity of means does not mean simplicity of outcomes. Great conceptual depth of understanding can be achieved through "simple" inquiry activities. Remember, however, that the purpose of such activities is to serve as starting points for students to assimilate and/or accommodate a coherent "science story" in light of what they already know and to revise what they believe to be true that is, in fact, mistaken.

3. Economical

NSES Program Standard D states that the "K–12 science program must give students access to appropriate and sufficient resources, including quality teachers, time, materials, and equipment, adequate and safe space, and the community" (NRC 1996, p. 218). Learning activities that require specialized lab facilities and supplies and/or time-intensive setup, use, and clean-up are unlikely to be regularly used by teachers with limited budgets and curricular and preparation time. Conversely, activities with simple material and setup requirements are more adaptable to standard classrooms across a range of grades. Especially useful activities are those that employ inexpensive, everyday materials that can be readily assembled, packaged (e.g., in zip-seal plastic bags of various sizes), re-supplied, and reused in other classes and school years. Household materials such as 2 L plastic soda bottles (De Vito 1995; Ingram 1993; Williams 1999), plastic tennis ball containers, 35 mm film canisters (De Vito 1993), balloons (Kaner 1993), and simple homemade or commercially available toys (O'Brien 1993; Sarquis, Williams, and Sarquis 1995; Taylor, Poth, and Portman 1995) can be used to heighten student interest in science. Teaching science with everyday materials and simple inexpensive devices not only saves money and time but also conveys that science happens all around us in our everyday life experiences (Sae 1996). Also, because using low-cost household supplies saves money, the available science budget can be stretched to purchase materials that might otherwise be considered too expensive (e.g., Activity #18/handheld generators, and Activity #21/polymer balls).

4. Enjoyable

NSES Teaching Standard A calls for teachers to "meet the interests, knowledge, understanding, abilities, and experiences of students" (NRC 1996, p. 30). Effective science teachers establish classroom environments where students come to class with anticipation and leave with regret. To achieve this objective, science teachers have all of nature's wonders to use as props in conjunction with other more generic teaching skills (Tauber and Sargent Mester 2007). Learning requires *mental* effort by the learner and, not uncommonly, some fruitful frustration. But learning should also typically involve fun, both in discovering and mastering content and skills.

Humans are neurologically wired to learn, and emotions are a key factor in cognition (Damasio 1994; Goleman 1995; McCombs and Whisler 1997). Our emotions serve as a kind of red-yellow-green traffic-light system that helps us rapidly assess how much and what kind of cognitive processing is needed in a given situation. Cognitive psychologist Howard Gardner states that idea in this way:

> Emotions serve as an early warning system, signaling topics and experiences that students find pleasurable to engage in, as well as those that may be troubling, mystifying or off-putting. Creating an educational environment in which pleasure, stimulation, and challenge flourish is an important mission. Also, students are more likely to learn, remember, and make subsequent use of those experiences with respect to which they had strong—and one hopes, positive—emotional reactions…. Any portrait of human nature that ignores motivation and emotion proves of limited

use in facilitating human learning and pedagogy…. If one wants something to be attended to, mastered, and subsequently used, one must wrap it in a context that engages the emotions. (Gardner 1999, p. 7)

Both discrepant-event approaches and quotes from famous scientists (as provided in this book) put a human face on otherwise abstract science concepts. Enjoyable science activities help students focus their mental energy by amplifying attention to relevant phenomena and by suppressing irrelevant distractions, thereby motivating them to stay tuned to the learning channel. Similarly, the often humorous science education analogies associated with each discrepant-event activity in this book combine an emotional and cognitive connection for teachers when they experience the same activities in a professional development context. Whatever the age of a learner, motivational and affective factors are critical to activating his or her attention and catalyzing cognitive processing (APA 1997).

5. Effective

NSES Program Standard B states that the "program of study in science for all students should … emphasize student *understanding through inquiry*…" (NRC 1996, p. 212). The fact that a given inquiry activity "works" is important but not sufficient to ensure that the activity will win out over other competitors for students' attention or that it will lead to improved conceptual understanding. Classroom science learning is a participatory sport that depends on lively interactions between a learner and "live" phenomena, other learners, a teacher, a textbook, and other mediating technologies. In too many cases, it is not so much true that students fail science but rather that science (instruction) fails them because it does not excite their interests and efforts. If we truly are to provide "science for all Americans," teachers must give their students systematic, intentional exposure to multiple representations of a concept using different senses, multiple intelligences (Armstrong 2000), and appropriately sequenced and scaffolded learning activities (Donovan and Bransford 2005; NRC 2007; Michaels et al. 2008). In addition, instructional units that use curriculum-embedded assessments will increase attention, cognitive processing, retention, and retrieval and transfer of knowledge (Bransford, Brown, and Cocking 1999; NRC 2007; Singer, Hilton, and Schweingruber 2006).

NSES Teaching Standard B describes effective teachers as those who

> guide and facilitate learning…; focus and support inquiries while interacting with students; orchestrate discourse among students about scientific ideas; challenge students to accept responsibility for their own learning; recognize and respond to student diversity and encourage all students to participate fully in science learning; [and] encourage and model the skills of scientific inquiry, as well as curiosity, openness to new ideas and data, and skepticism that characterize science. (NRC 1996, p. 32)

Finding the happy medium between a task that is too difficult for students (resulting in frustration and anxiety) and one that is too easy (leading to boredom) is part of the science and art of teaching (Bransford, Brown, and Cocking 1999; Brooks and Brooks 1999). Learners should be presented with optimally targeted and tailored challenges (see Activity #4). Effective teachers engage students in the students' zones of proximal development (ZPD)—that is, where cognitive and emotional challenges are balanced with comfort and a willingness to "stretch into the unknown" with the support of a competent teacher (Vygotsky 1978). Being "in the zone" creates a self-reinforcing, positive feedback loop and a corresponding feeling of "flow" (Csikszentmihalyi 1990). The flow leads to optimized cognitive efficiency and increases intrinsic motivation and resiliency for facing future challenges (Goleman 2006).

Part of the challenge in teaching science is that, in a number of ways, it requires "uncommon sense" and "unnatural" ways of thinking (Cromer 1993; Wolpert 1992). This is evident from the fact that the same tenacious student misconceptions will be unearthed whether one is working with upper-elementary, middle school, or high school students (Duit 2009; Kind 2004; Meaningful Learning Research Group; Olenick 2008; Operation Physics; Science Hobbyist). Equally important from the teacher-as-learner perspective, "it is patently incorrect to assume that the ability to teach the nature of science comes naturally" (Lederman and Neiss 1997 p. 2). Each section in each activity in this book is designed to help teachers do some pedagogical "stretching" and bridge the gap between everyday, commonsense thinking and scientific ways of observing, reasoning, and talking.

Discrepant-event activities—that is, interactive, inquiry-oriented demonstrations and experiments with counterintuitive and/or otherwise surprising outcomes—have been part of science education since at least the time of Michael Faraday's popular public lectures in the mid-1800s. Festinger (1957) and Piaget (1973) emphasized the importance of engaging students with learning experiences that activate attention and create cognitive conflict, dissonance, or disequilibrium in regard to their prior, more limited, or erroneous conceptions. Discrepant (or anomalous) events can be seen as "irritants" that bring about conceptual change, a process that is analogous to the way an oyster transforms a particle of sand into a pearl.

Oysters respond automatically. Learners, however, when presented with an anomaly, have a variety of conscious as well as unconscious ways to preserve their old ideas.

1. They may fail to notice the anomaly through inattention. This is why dramatic discrepant-event activities are so pedagogically powerful.

2. They may notice the anomaly but see it as a special case or magic trick and may therefore decide to hold onto their preexisting theories. Multiple discrepant-event activities are sometimes needed to dislodge and replace tenacious misconceptions.

3. They may make minor modifications to their preexisting theories as a way of accounting for the anomaly. This is why it is important for teachers to engage students with additional empirical evidence, logical arguments, and skeptical review—all of which present the correct scientific theory as being more intelligible, plausible and fruitful than their old ideas. (Posner et al. 1982)

Appropriately targeted, anomaly-based conceptual challenges motivate a need-to-notice-and-know that can be addressed by using explorations and guided discussions to help students construct new, improved (or substantially refined) conceptual schemata (Chinn and Brewer 1993; Friedl and Koontz 2005; Gabel 1995; Hestenes; Mintzes, Wandersee, and Novak 1998; NRC 2007; Stepans 1996). Appendix B describes the 5E Teaching Cycle that incorporates the use of discrepant-event activities into a larger unit-level approach that integrates (rather than separates) curricular planning, instruction, and assessment (see Appendix B and Bybee et al. 2006).

Empirical support for the power of discrepant events also comes from researchers who study children's "naive science" (or alternative, preinstructional misconceptions) by using live, multimedia-based, or paper-and-pencil forms of discrepant events in their diagnostic and formative assessments and clinical interviews (e.g., Driver, Guesne, and Tiberghein 1985; Driver et al. 1994; Fensham, Gunstone, and White 1994; Keeley, Eberle, and Farrin 2005; Keeley, Eberle, and Tugel 2007; Osborne and Freyberg 1985; Treagust, Duit, and Fraser 1996; White and Gunstone 1992). Predict-observe-explain (POE), minds-on demonstrations have been cited as important complements to hands-on laboratory experiences because they encourage the manipulation of ideas (Hofstein and Lunetta 2004). Similarly, one of the major instructional recommendations of the book *How Students Learn Science in the Classroom* is to "provide opportunities for students to experience discrepant events that allow them to come to terms with the shortcomings in their everyday models" (Donovan and Bransford 2005, p. 571; see also Hestenes). An effective science teacher uses discrepant-event activities to do much more than excite interest and entertain (McCormack 1990); he or she uses these activities to help students see the need for and logic and economy of new conceptual, explanatory models.

Effective teachers also empower learners to draw on valid prior understandings and to develop skills to become more independent learners. Historically, ancient myths, fairy tales, and religious texts—and master teachers as well—have tapped into the human brain's natural propensity to use **analogies** as cognitive tools to build bridges of understanding between the familiar and the unfamiliar. Indeed, it has been shown that both explicit (intentional) and implicit (unconscious) use of metaphors and analogies is pervasive in everyday oral and written communication (Lakeoff and Johnson 1980). Analogical reasoning is especially critical when dealing with abstract or nonobservable concepts and complex processes within systems with multiple interacting parts. The "big ideas," core concepts, and theories of science (e.g., atoms and molecules, cells and evolution, plate tectonics, and conservation of matter-energy) took generations of development to reach their current forms. Unfortunately, textbooks often expect "super-students to leap the tall buildings" of such abstract ideas "in a single bound." Both research- and practitioner-oriented publications recommend that teachers use bridging

and metaphorical analogies as cognitive scaffolding to help students work within their zone of proximal development to reconstruct, retain, retrieve, and refine scientifically accurate explanatory models (Camp and Clement 1994; Dagher 1998; Duit 1991; Gabel 1995; Gilbert and Ireton 2003; Glynn 1991; Glynn and Takahashi 1998; Hackney and Wandersee 2002; Harrison and Coll 2008; Hoagland and Dodson 1995; Lawson 1993; Marzano, Pickering, and Pollack 2001; NRC 2000, 2007; Packard 1994).

It is important to acknowledge that when teachers and students use new instructional strategies, they are pushed out of their comfort zones, and short-term "implementation dips" are likely to occur (Hughes et al. 2009). Learning how to engage students in an interactive "narrative of discovery" rather than presenting a monologue based on the "rhetoric of conclusions" (DeBoer 1991) takes time and guided practice on the part of both teachers and students (Liem 1990; Llewellyn 2005). Both types of learners need strong, regular encouragement. In addition, effectiveness of new ways of teaching and learning must not be judged prematurely (in time) or narrowly (with respect to immediate, cognitive-only assessment outcomes). Both teachers and students need time to learn how to teach and to learn in new ways.

6. Relevant

NSES Program Standard B states that the "program of study in science should be developmentally appropriate, interesting, and relevant to students' lives" (NRC 1996, p. 212). As we know, an effective teacher does not assemble a series of neat tricks in a random sequence without thinking about their relevance to a specific audience and to the broader society. The issue of relevance raises these questions: For whom are lessons intended? For what purposes are they created? Who decides what is relevant to a curriculum? As such, the issue of relevance elicits a multiplicity of answers from both research and policy arenas (Aikenhead 2006). Both the AAAS Benchmarks (1993) and the National Science Education Standards (NRC 1996) highlight the need to teach the big ideas or conceptual themes of science as well as the habits of mind that prepare diverse learners for their lives as skilled, adaptable workers and informed citizens. Part of the role of a teacher and curriculum is to connect with and enlarge students' preinstructional, limited sense of what is relevant. This book's activities are designed to pose problems of emerging relevance (Brooks and Brooks 1999) that get teachers and their students to think both within and outside the box of their prior knowledge and to see big picture connections.

Relevance also requires that learning activities be part of integrated instructional units (NSTA 2007b; Singer, Hilton, and Schweingruber 2006) and multiyear learning progressions (Michaels, Shouse, and Schweingruber 2008; NRC 2007). Given this book's multiple audiences—grades 5–12 science teachers, science educators, and professional development specialists—the activities are *not* presented as science units but rather are grouped by science education themes as related to the underlying principles of cognitive learning theory—an approach, one hopes, that is most relevant for science teachers.

The 5E Teaching Cycle: An Integrated Curriculum-Instruction-Assessment Model

Cognitive science research argues that effective instruction is much more than a series of lessons that contain periodic entertaining surprises. What is needed is instruction by the teacher that promotes student construction of progressively more refined scientific models. For that to occur, individual lessons need to be sequenced in such a way that the whole (unit) is greater than the sum of its parts. Conventionally, teaching is simplistically conceptualized as a linear, noniterative sequence with three separate steps: (1) design the curriculum (or "plan"), (2) implement instruction (or "teach"), and (3) assess the students (or "test" and "grade"). In contrast, more "intelligent" science units integrate those three steps into iterative cycles that scaffold students' ever-evolving understanding of both science content and the nature of science.

The 5E Teaching Cycle is a research-informed *integrated instructional unit* design model that was developed by the Biological Sciences Curriculum Study (BSCS) group. BSCS has been developing inquiry-oriented textbooks for over 40 years. In the late 1980s, BSCS modified an earlier three-phase learning cycle to create the 5E Teaching Cycle (Engage, Explore, Explain, Elaborate, Evaluate). The 5E model has much to offer classroom teachers whether or not they use BSCS-designed curricula (Bybee et al. 2006). Although individual lessons may be conceived of as "mini-5Es," this instructional model is especially powerful when used to frame units of one to two weeks (with one or more days devoted to each "E").

When this book is used to teach grades 5–12 science, the discrepant-event demonstrations are especially useful as Engage-phase activities. The hands-on explorations can be used during the Engage phase or they can be modified to become more formal Explore-phase laboratory investigations. Alternatively, depending on the instructional sequence chosen by the teacher—and on adaptations he or she has made—either type of activity (discrepant event or laboratory investigations) might fit into the Explain, Elaborate, or Evaluation phase. I provide the following descriptions of each of the five phases to give teachers a better sense of how they can creatively use this book's activities with their grades 5–12 students. In that process, teachers will be challenged to reconsider some of their *own* preexisting ideas about science content and science teaching.

Engage

The opening act of the 5E Teaching Cycle—Engage—can be thought of as the "golden grabber" or sales pitch whereby students are enticed to allow the teacher into the partially opened doors of their minds. Once inside, the teacher-seller asks students to consider whether their old mental models explain the just-introduced phenomenon, or whether, perhaps, they might need to upgrade to a more efficient, scientifically valid model.

The purposes of this phase are to (1) create a need-to-know that motivates students to inquiry about, and form their own opinions about, the relevance of the unit's topic(s) and (2) activate and diagnostically assess students' prior conceptions (including misconceptions) as a prerequisite step to begin constructing (or reconstructing) their mental models. Given the conservatism (or inertia) of the human mind, both cognitive dissonance (minds-on) and affective or emotional engagement (hearts-on) are needed to address the students' spoken or unspoken questions: "So what?" and "Who cares?" Discrepant-event demonstrations, simple hands-on explorations (that ask students to predict-observe-explain), visually surprising video clips, engaging stories with a "twist," and preinstructional assessments (e.g., agree-disagree-unsure questionnaires) are all good ways to raise questions that warrant subsequent exploration by the class.

During the Engage phase, it is important to avoid introducing and defining many scientific terms and to avoid a teaching-as-telling approach. Let students struggle with the questions elicited by the activities; answers will come via hands-on and minds-on experiences in subsequent phases. Science class should not be a game of "How many did you get right?" but rather a place where "miss-takes" are seen as steps in the right direction and question generation and refinement are as important as the answers. Challenge students to consider multiple plausible explanations and avoid leading them to premature closure on the answer.

To expand students' scientific visions, teachers should make the familiar world of everyday objects and observations seem "strange" and the strange world of scientific equipment, procedures, and concepts seem more familiar. *Spending at least one full lesson at the start of each unit in the Engage phase is not excessive.* If students fail to buy the teacher's sales pitch or if instruction starts with an inaccurate view of students' prior conceptions and skills, time will be wasted trying to force feed students "food" they think they don't like, don't see the need to eat, and are unable to digest.

Explore

In the Explore phase, students participate in minds-on, scientific play in the form of guided-inquiry investigations that extend the common experiential base provided in the Engage phase. It is not a time for students to verify answers given to them by the teacher or textbook. In fact, the teacher may need to modify published laboratory exercises to eliminate introductory statements that give away the answers without first having students develop and/or explore the questions.

In the Explore phase, more than during any other phase, the teacher plays the role of the guide on the side, not the sage on the stage. Students typically work in cooperative learning teams where they manipulate objects and/or simple systems. Complex, dangerous, very large or very small scale, or otherwise inaccessible systems may be explored via manipulative models, participatory simulations (e.g., student-molecules simulating the kinetic molecular theory and student-predators attacking prey), and computer animations (as found in the Internet Connections that are part of each activity in this book).

The Explore phase may also involve the introduction or refinement of science process skills, such as the use of data analysis techniques and equipment to make quantitative measurements. Hands-on, multisensory experiences should lead students to develop refined, minds-on questions (and tentative answers) about both the Explore investigations and the initial Engage-phase activities.

Because science is based on empirical evidence, logical arguments, and skeptical review (rather than simple faith in authority), students should keep notebooks of observations, inferences, experimental results, questions raised, and conclusions drawn from their experiences. Teacher modeling and testing of student lab skills and active monitoring of student learning by asking probing questions are essential to ensure that hands-on time leads to minds-on learning. Once again, it is essential to avoid the pedagogical errors of providing premature answers or step-by-step solutions to students' questions; encourage students to be self-reliant and interdependent with their cooperative learning partners.

Explain

The Explain phase is the appropriate time for teachers to formally introduce scientific concepts, principles, and terminology that guide students to construct logical explanatory arguments for the empirical evidence gathered from previous demonstrations and experiments. In contrast to the conventional giving-the-answers approach, teachers use probing Socratic questioning, additional demonstrations, guided-inquiry experiments, and interactive multimedia. Teachers also take time to help students resolve unanswered questions. This is a good time to encourage students to skeptically review any remaining misconceptions.

Collaboratively developed analogies and concept maps (and other graphic organizers) are powerful ways of building conceptual bridges between the students' valid preconceptions and new ideas—to help them see the "big picture," that is, the conceptual forest in place of individual phenomological trees. The teacher emphasizes conceptually rich meaning-making rather than short-term memorization of isolated facts intended for regurgitation on low-level tests. This is also an opportune time to feature interdisciplinary connections, including reading, writing, and mathematics skills as linked to the students' science notebooks, the textbook, problem sets, and websites.

Elaborate

In the Elaborate phase, the teacher—using formative assessment—informally "tests" the students' understanding by challenging students to apply core concepts and theories in new or less-familiar contexts. Such assessment may involve the use of real-world applications or problems; teacher demonstrations of new but related phenomena; open-ended, hands-on investigations; and library or internet investigations. The objective is to encourage the students to take ownership of, and become confident with, the underlying science. Also, those students who have not yet quite "got it" have another opportunity to have an "Aha!" experience.

The teacher can introduce related concepts as extensions of the core concepts (especially for the more advanced students) or as bridges to a subsequent cycle. In any case, students gain both competence and confidence in their newly acquired understanding as they are challenged to use higher-order thinking skills (based on the top three levels of Bloom's taxonomy: analysis, synthesis, and evaluation).

Students' performances on subsequent high-stakes summative assessments (as discussed in the next section) will more accurately reflect their understanding of the targeted concepts if they've had scaffolded encounters with low-stakes formative assessments where their "misstakes" can corrected and the teacher can review topics as needed.

Evaluate

Curriculum-embedded diagnostic and formative assessments are used throughout the first four phases. The Evaluate phase is the time for

summative assessment. This type of assessment is typically associated with "grading" individual students on timed, objective, paper-and-pencil tests. Other types of summative assessment are also available. For example, the teacher can challenge students to predict-observe-explain new discrepant-event demonstrations or explain the science behind cartoons related to the unit concepts. Other alternative summative assessments include independent laboratory or field-based investigations; science-technology-society debates; letters to the editor; oral and/or written reports; multimedia projects; models or hallway displays; and "science is magical but not magic" presentations to younger students.

When combined, conventional and alternative assessments should be used to determine the following:

- The extent to which students' valid preconceptions have been extended, or, if those preconceptions were faulty, how much students' understanding has moved toward more scientifically accurate conceptions (e.g., if a written diagnostic test was used during the Engage phase, some of those items can be repeated on a unit test to check for gain scores).
- The extent to which students have progressed in their abilities to use scientific habits of mind and inquiry skills (e.g., framing and addressing questions in light of empirical evidence, logical argument, and skeptical review).
- Whether student interest in the topic has increased and whether attitudes about the relevance of science have become more positive. Additionally, remaining misconceptions and unresolved questions may suggest that review is needed before proceeding to the next unit.

One risk of any "neat" science activity book is that the activities can be used as magic tricks to increase students' interest in units that otherwise rely too heavily on teaching as telling and learning as listening. This book's dual focus—on discrepant-event activities and on science education visual participatory analogies—is designed to *engage* teachers in the *exploration* of the science and the art of research-informed, best practice teaching. Each activity contains *explanations* and *elaborations* related to the domains of both science content and pedagogical principles. Ultimately, of course, the only meaningful *evaluation* of "lessons learned" occurs in each science teacher's classroom.

Science Content Topics*

Air and Air Pressure

7. Identification Detectives: Sounds and Smells of Science

8. Two-Balloon Balancing Act: Constructivist Teaching

13. Sound Tube Toys: The Importance of Varying Stimuli

16. Air Mass Matters: Creating a Need-to-Know

27. Invisible Gases Matter: Knowledge Pours Poorly

30. Tornado in a Bottle: The Vortex of Teaching and Learning

Biological Applications and Analogies

3. Burning a Candle at Both Ends: Classrooms as Complex Systems (Extension #3/ cellular espiration)

4. Perceptual Paradoxes: Multisensory Science and Measurement (sensory adaptations and survival)

5. Optical Illusions: Seeing and Cognitive Construction (perception and cognition)

6. Utensil Music: Teaching Sound Science (see Debriefing/sensory variations in species)

7. Identification Detectives: Sounds and Smells of Science (sensory adaptations and survival)

8. Two-Balloon Balancing Act: Contructivist Teaching (Extension #2/aneurisms in adults and alveoli collapse in premature babies)

*All topics are identified by activity number and activity name.

Density (including Buoyancy and Heat/Convection)

Inquiry in Teaching and Learning

Law of Conservation of Energy

Light, Optics, Optical Illusions, and Mirrors

Magnetism

Measurement Skills

Phase Changes

Research Cited

Aicken, F. 1991. *The nature of science.* 2nd ed. Portsmouth, NH: Heinemann.

Aikenhead, G. S. 2006. *Science education for everyday life: Evidence-based practice.* New York: Teachers College Press.

American Association for the Advancement of Science (AAAS). 1989. *Science for all Americans.* New York: Oxford University Press.

American Association for the Advancement of Science (AAAS). 1993. *Benchmarks for science literacy.* New York: Oxford University Press. *http://project2061. aaas.org*

American Association for the Advancement of Science (AAAS). 2001. *Atlas of science literacy.* Vol. 1. Washington, DC: AAAS and Arlington, VA: NSTA Press.

American Association for the Advancement of Science (AAAS). 2007. *Atlas of science literacy.* Vol. 2. Washington, DC: AAAS and Arlington, VA: NSTA Press.

American Psychological Association (APA). 1997. *Learner-centered psychological principles: A framework for school redesign and reform.* Washington, DC: APA Center for Psychology in Schools and Education. *www.apa.org/ed/cpse/LCPP.pdf*

Anderson, L. W. and D. R. Krathwohl, eds. 2001. *A taxonomy for learning, teaching, and assessing: A revision of Bloom's taxonomy of educational objectives.* New York: Addison Wesley Longman.

Armstrong, T. 2000. *Multiple intelligences in the classroom.* Alexandria, VA: Association for Supervision and Curriculum Development.

Asimov, I. 1976. *Isaac Asimov's biographical encyclopedia of science & technology.* New York: Avon Books.

Banilower, E. R., S. E. Boyd, J. D. Pasley, and I. R. Weiss. 2006. *Lessons from a decade of mathematics and science reform: The local systemic change through teacher enhancement initiative.* Chapel Hill, NC: Horizon Research. *www.pdmathsci.net/findings/report/32*

Bloom, B. S., ed., and M. D. Engelhart, E. J. Furst, W. H. Hill, and D. R. Krathwohl. 1956. *Taxonomy of educational objectives: Handbook I; Cognitive domain.* New York: David McKay.

Bloomfield, L. A. 2007. *How everything works: Making physics out of the ordinary.* Hoboken, NJ: John Wiley and Sons.

Bower, B. 1992. Brother Stroop's enduring effect: A mental task devised nearly 60 years ago still intrigues psychologists. *Science News* 141 (May): 312–314.

Research Cited

Bradley, A. 2004. *Your memory: A user's guide*. Richmond Hill, Ontario: Firefly Books. (This book and Higbee 2001 have the strongest research bases for the demonstrations in Activity #12.)

Bransford, J. D., A. L. Brown, and R. R. Cocking, eds. 2000. *How people learn: Brain, mind, experience, and school*. Washington, DC: National Academies Press.

Brooks, J. G., and M. G. Brooks. 1999. *In search of understanding: The case for constructivist classrooms*. Alexandria, VA: Association for Supervision and Curriculum Development.

Buzan, T. 1991. *Use your perfect memory: Dramatic new techniques for improving your memory*. 3rd ed. New York: Plume.

Bybee, R. W., ed. 2002. *Learning science and the science of learning*. Arlington, VA: NSTA Press.

Bybee, R. W., J. A. Taylor, A. Gardner, P. Van Scotter, J. Carlson Powell, A. Westbrook, and N. Landes. 2006. *BSCS 5E Instructional model: Origins, effectiveness and applications*. Colorado Springs, CO: BSCS. *www.bscs.org/pdf/5EFull Report.pdf* (65 pages) and *http://bscs.org/pdf/bscs5eexecsummary.pdf* (19 pages)

Camp, C. W., and J. J. Clement. 1994. *Preconceptions in mechanics: Lessons dealing with students' conceptual difficulties*. Dubuque, IA: Kendall/Hunt.

Chinn, C. A., and W. F. Brewer. 1993. The role of anomalous data in knowledge acquisition: A theoretical framework and implications for science instruction. *Review of Educational Research* 63 (1): 1–49.

Cochran, K. F. 1997. Pedagogical content knowledge: Teacher's integration of subject matter, pedagogy, students, and learning environments. Brief. *Research Matters to the Science Teacher*. No. 9702. National Association of Research in Science Teaching. *www.narst.org/publications/research.cfm*

Cocking, R. R., J. P. Mestre, and A. L. Brown, eds. 2000. New developments in the science of learning: Using research to help students learn science and mathematics. *Journal of Applied Developmental Psychology, Special Issue* 21(1).

Cromer, A. 1993. *Uncommon sense: The heretical nature of science*. New York: Oxford University Press.

Csikszentmihalyi, M. 1990. *Flow: The psychology of optimal experience*. New York: Harper and Row.

Dagher, Z. 1998. The case for analogies in teaching science for understanding. In *Teaching science for understanding: A human constructivist view*, ed. J. J. Mintzes, J. H. Wandersee, and J. D. Novak. San Diego, CA: Academic Press.

Damasio, A. R. 1994. *Descartes' error: Emotion, reason, and the human brain*. New York: Grosset/Putnam Book.

DeBoer, G. E. 1991. *A history of ideas in science education: Implications for practice*. New York: Teachers College Press.

Devito, A. 1982. *Teaching with eggs*. West Lafayette, IN: Creative Ventures.

Devito, A. 1993. *Recycling 35mm canisters for the teaching of science*. West Lafayette, IN: Creative Ventures.

Devito, A. 1995. *Recycling two-liter containers for the teaching of science*. West Lafayette, IN: Creative Ventures.

Donovan, M. S., J. D. Bransford, and J. W. Pelligrino, eds. 1999. *How people learn: Bridging research and practice.* Washington, DC: National Academy Press.

Donovan, M. S., and J. D. Bransford, eds. 2005. *How students learn: Science in the classroom.* Washington, DC: National Academies Press.

Doran, R.L. February 1974. Hammer or sponge? *The Science Teacher* 41 (2): 34–35.

Driver, R., E. Guesne, and A. Tiberghein, eds. 1985. *Children's ideas in science* London: Open University Press.

Driver, R., A. Squires, P. Rushworth, and V. Wood-Robinson. 1994. *Making sense of secondary science: Research into children's ideas.* London: Routledge.

Duit, R. 1991. On the role of analogies and metaphors in learning science. *Science Education* 75: 649–672.

Duit, R. 2009. *Bibliography of students' and teachers' conceptions & science education.* Kiel, Germany: Institute for Science Education at the University of Kiel. *www.ipn.uni-kiel.de/aktuell/stcse/stcse.html*

Engelhart, M. D., E. J. Furst, W. H. Hill, and D. R. Krathwohl. 1956. *Taxonomy of educational objectives: Handbook I; Cognitive domain,* ed. B. S. Bloom. New York: David McKay Co.

Fensham, P., R. Gunstone, and R. White, eds. 1994. *The content of science: A constructivist approach to its teaching and learning.* London: Falmer Press.

Ferguson, R. F., and H. F. Ladd. 1996. *Holding schools accountable: Performance-based reform in education.* Washington, DC: Brookings Institution.

Festinger, L. 1957. *A theory of cognitive dissonance.* Stanford, CA: Stanford University Press.

Flinn Scientific, Inc. Safety resources. *www.flinnsci.com/Sections/Safety/safety.asp*

Friedl, A. E., and T. Y. Koontz. 2005. *Teaching science to children: An inquiry approach.* 6th ed. New York: McGraw Hill.

Gabel, D. 1995. Improving student achievement in science. In *Handbook of research on improving student achievement,* ed. G. Cawelti, Chapter 9. Arlington, VA: Educational Research Service.

Gardner, H. 1999. *The disciplined mind: What all students should understand.* New York: Simon and Schuster.

Gilbert, S. W., and S. Watt Ireton. 2003. In *Understanding models in Earth and space science.* Arlington, VA: NSTA Press. (See especially Chapter 2, "Similes, Analogies, and Metaphors.")

Gilovich, T. 1991. *How we know what isn't so: The fallibility of human reason in everyday Life.* New York: Free Press.

Glynn, S. M. 1991. Explaining science concepts: A teaching-with-analogies model. In *The psychology of learning science,* ed. S. Glynn, R. Yeany, and S. Britton, 219–240. Hillsdale, NJ: Lawrence Erlbaum. See also Dr. Glynn's website: *www.coe.uga.edu/twa*

Glynn, S. M., and T. Takahashi. 1998. Learning from analogy-enhanced science text. *Journal of Research in Science Teaching* 35 (10): 1129–1149.

Goleman, D. 1995. *Emotional intelligence: Why it can matter more than IQ.* New York: Bantam Books.

Research Cited

Goleman, D. 2006. *Social intelligence: The new science of human relationships*. New York: Bantam Books.

Good, R. G., J. D. Novak, and J. H. Wandersee, eds. 1990. Perspectives on concept mapping. *Journal of Research in Science Teaching, Special Issue* 27 (10). *www.narst.org*

Grant, J. 2006. *Discarded science: Ideas that seemed good at the time*. Wisely, Surrey, UK: Facts, Figures, and Fun.

Gribbin, J. 2002. *The scientists: A history of science told through the lives of its greatest inventors*. New York: Random House.

Hackney, M. W., and J. H. Wandersee. 2002. *The power of analogy: Teaching biology with relevant classroom-tested activities*. Reston, VA: National Association of Biology Teachers.

Hakim, J. *The story of science* (series): *Aristotle leads the way* 2004. *Newton at the center* 2005. *Einstein adds a new dimension* 2007. Washington, DC: Smithsonian Books and Arlington, VA: NSTA Press.

Harrison, A. G., and R. K. Coll, eds. 2008. *Using analogies in middle and secondary science classrooms: The FAR guide—An interesting way to teach with analogies*. Thousand Oaks, CA: Corwin Press.

Harrison, A. G., R. K. Coll, and D. Treagust. 1994. Science analogies. *The Science Teacher* (April): 40–43.

Hart, L. 1983. *Human brain, human learning*. New York: Longman.

Harvard-Smithsonian Center for Astrophysics, Science Education Department. MOSART: Misconception Oriented Standards-based Assessment Resource for Teachers: *www.cfa.harvard.edu/sed/projects/mosart.html*

Hellemans, A., and B. Bunch. 1988. *The timetables of science: A chronology of the most important people and events in the history of science*. New York: Simon and Schuster.

Hestenes, D. Modeling instruction in high school physics. *http://modeling.asu.edu/modeling-HS.html*

Higbee, K. 2001. *Your memory: How it works and how to improve it*. 2nd rev. ed. New York: Marlowe and Co. (This book and Bradley 2004 have the strongest research bases for the demonstrations in Activity #12.)

Hoagland, M., and B. Dodson. 1995. *The way life works*. New York: Times Books/Random House.

Hofstein, A., and V. N. Lunetta. 2004. The laboratory in science education: Foundations for the twenty-first century. *Science Education* 88:28–54.

Hughes, R. L., R. C. Ginnett, and G. J. Curphy. 2009. *Leadership: Enhancing the lessons of experience*. 6th ed. Chicago: McGraw-Hill/Irwin.

Ingram, M. 1993. *Bottle biology: An idea book for exploring the world through soda bottles and other recyclable materials*. Dubuque, IA: Kendall/Hunt.

Jensen, E. 1998. *Teaching with the brain in mind*. Alexandria, VA: Association for Supervision and Curriculum Development.

Kaner, E. 1993. *Balloon science: A science book bursting with more than 50 balloon experiments and activities*. Reading, MA: Addison-Wesley.

Keeley, P., F. Eberle, and L. Farrin. 2005. *Uncovering student ideas in science (vol. 1): 25 formative assessment probes*. Arlington, VA: NSTA Press.

Keeley, P., F. Eberle, and J. Tugel. 2007. *Uncovering student ideas in science (vol. 2): 25 more formative assessment probes.* Arlington, VA: NSTA Press.

Kind, V. 2004. Beyond appearance: Students' misconceptions about basic chemical ideas. *www.rsc.org/education/teachers/learnnet/pdf/LearnNet/rsc/miscon.pdf*

Kuhn T. 1962. *The structure of scientific revolutions.* Chicago: University of Chicago Press.

Kwan, T., and J. Texley. 2002. *Exploring safely: A guide for elementary teachers.* Arlington, VA: NSTA Press.

Kwan, T., and J. Texley. 2003. *Exploring safely: A guide for middle school teachers.* Arlington, VA: NSTA Press.

Lakeoff, G., and M. Johnson. 1980. *Metaphors we live by.* Chicago: University of Chicago Press.

Lawson, A. E., ed. 1993. The role of analogy in science and science teaching. *Journal of Research in Science Teaching, Special Issue* 30(10). *www.narst.org*

Lederman, N. G. 1992. Students' and teachers' conceptions of the nature of science: A review of the research. *Journal of Research in Science Teaching* 29:331–359.

Lederman, N. G. 1999. Teachers' understanding of the NOS and classroom practice: Factors that facilitate or impede the relationship. *Journal of Research in Science Teaching* 36(8): 916–929.

Lederman, N. G., and M. L. Neiss. 1997. The nature of science: Naturally? *School Science and Mathematics* 97: 1–2.

Liem, T. L. 1990. *Invitations to science inquiry.* 2nd ed. Chino Hills, CA: Science Inquiry Enterprise.

Llewellyn, D. 2005. *Teaching high school science through inquiry.* Thousand Oaks, CA: Corwin Press.

Lorayne, H. 1990. *Super memory—super student: How to raise your grades in 30 days.* Boston: Little, Brown.

Lorayne, H., and J. Lucas. 1996. *The memory book: The classic guide to improving your memory at work, at school, and at play.* New York: Ballantine Books.

Lortie, D. C. 1975. *Schoolteacher: A sociological study.* Chicago: University of Chicago Press.

Loucks-Horsley, S., P. W. Hewson, N. Love, and K. E. Stiles. 1998. *Designing professional development for teachers of science and mathematics.* Thousand Oaks, CA: Corwin Press.

MacLeod, C. M. 1991. Half a century of research on the Stroop effect: An integrative approach. *Psychological Bulletin* 109 (2): 163–203.

Marzano, R. L., D. J. Pickering, and J. E. Pollock. 2001. Identifying similarities and differences. In *Classroom instruction that works: Research-based strategies for increasing student achievement*, Chapter 2. Alexandria, VA: Association for Supervision and Curriculum Development.

Matthews, M. R. 1994. *Science teaching: The role of history and philosophy of science.* New York: Routledge.

McComas, W. F. 1996. Myths of science: Reexamining what we know about the nature of science. *School Science and Mathematics* 96:10–16.

McCombs, B. L., and J. S. Whisler. 1997. *The learner-centered classroom and school: Strategies for increasing student motivation and achievement.* San Francisco: CA: Jossey-Bass.

Research Cited

McCormack, A. J. 1990. *Magic and showmanship for teachers*. Riverview, FL: Idea Factory.

Meaningful Learning Research Group. Misconceptions conference proceedings. *www2.ucsc.edu/mlrg/mlrgarticles.html*

Michael, J. A., and H. I. Modell. 2003. *Active learning in secondary and college science classrooms: A working model for helping the learner to learn*. Mahwah, NJ: Lawrence Erlbaum.

Michaels, S., A. W. Shouse, and H. A Schweingruber. 2008. *Ready, set, science! Putting research to work in K-8 science classrooms*. Washington, DC: National Academies Press. *www.nap.edu/catalog.php?record_id=11882*

Miller, G. A. 1956. The magical number seven, plus or minus two: Some limits on our capacity for processing information. *The Psychological Review* 63:81–97.

Mintzes, J. J., J. H. Wandersee, and J. D. Novak, eds. 1998. *Teaching science for understanding: A human constructivist view*. New York: Academic Press.

National Commission on Mathematics and Science Teaching (NCMST) for the 21st Century. 2000. *Before it's too late*. Washington, DC: Department of Education. *www.ed.gov/inits/Math/glenn/report.pdf*

National Commission on Teaching and America's Future (NCTAF). 1996. *What matters most: Teaching for America's future*. New York: NCTAF/Teachers College Press. *www.nctaf.org/documents/WhatMattersMost.pdf*

National Commission on Teaching and America's Future (NCTAF). 1997. *Doing what matters most: Investing in quality teaching*. New York: NCTAF/Teachers College Press. *www.nctaf.org/documents/DoingWhatMattersMost.pdf*

National Research Council (NRC). 1996. *National science education standards*. Washington, DC: National Academies Press. *www.nap.edu/catalog.php?record_id-4962*

National Research Council (NRC). 2000. *Inquiry and the national science education standards: A guide for teaching and learning*. Washington, DC: National Academies Press.

National Research Council (NRC). 2001a. *Educating teachers of science, mathematics, and technology: New practices for the new millennium*. Washington, DC: National Academies Press.

National Research Council (NRC). 2001b. *Classroom assessment and the national science education standards*. Washington, DC: National Academies Press.

National Research Council (NRC). 2007. *Taking science to school: Learning and teaching science in grades K-8*, ed. R. A. Duschl, H. A. Schweingruber, and A. W. Shouse. Washington, DC: National Academies Press. *www.nap.edu/catalog.php?record_id=11625*

National Science Board (NSB). 2006. *America's pressing challenge—Building a stronger foundation*. NSB 06-02. Washington, DC: National Science Foundation. *www.nsf.gov/statistics/nsb0602*

National Science Teachers Association (NSTA). 1990. The role of research in science teaching. NSTA Position Paper adopted in January 1990. Arlington, VA: NSTA. *www.nsta.org/about/positions/researchrole.aspx*

National Science Teachers Association (NSTA). 2000. The nature of science. NSTA Position Paper adopted in July 2000. Arlington, VA: NSTA. *www.nsta.org/about/positions/natureofscience.aspx*

National Science Teachers Association (NSTA). 2004. Science inquiry. NSTA Position Paper adopted in October 2004. Arlington, VA: NSTA. *www.nsta.org/about/positions/inquiry.aspx*

National Science Teachers Association (NSTA). 2006. Professional development in science education. NSTA Position Paper adopted in May 2006. Arlington, VA: NSTA. *www.nsta.org/about/positions/profdev.aspx*

National Science Teachers Association (NSTA). 2007a. Principles of professionalism for science educators. NSTA Position Paper adopted in June 2007. Arlington, VA: NSTA. *www.nsta.org/about/positions/professionalism.aspx.*

National Science Teachers Association (NSTA). 2007b. The integral role of laboratory investigations in science instruction. NSTA Position Paper adopted in February 2007. Arlington, VA: NSTA. *www.nsta.org/about/positions/laboratory.aspx*

National Science Teachers Association (NSTA). 2007c. Induction programs for the support and development of beginning teachers of science. NSTA Position Paper adopted in April 2007. Arlington, VA: NSTA. *www.nsta.org/about/positions/induction.aspx.*

National Science Teachers Association (NSTA). 2008. The role of e-learning in science education. NSTA Position Paper adopted in September 2008. Arlington, VA: NSTA. *www.nsta.org/about/positions/e-learning.aspx*

O'Brien, T. 1991. The science and art of demonstrations. *Journal of Chemical Education* 68:933–936.

O'Brien, T. 1992a. The concerns-based design and evaluation of an institute for chemical education demonstration workshop. *Science Educator* 1 (1): 21–26.

O'Brien, T. 1992b. Science in-service workshops that work for elementary teachers. *School Science and Mathematics* 92:422–426.

O'Brien, T. 1993. Teaching fundamental aspects of science toys. *School Science and Mathematics* 93:203–207.

O'Brien, T., C. Stannard, and A. Telesca. 1994. A baker's dozen of discrepantly dense demonstrations. *Science Scope* 18 (2): 35–38.

Olenick, R. P. 2008. Comprehensive conceptual curriculum for physics (C3P) project. Misconceptions and preconceptions in introductory physics. *http://phys.udallas.edu/C3P/Preconceptions.pdf*

Operation Physics. Children's misconceptions about science. *www.amasci.com/miscon/opphys.html*

Osborne, R., and P. Freyberg. 1985. *Learning in science: The implications of children's science.* London: Heinemann.

Packard, E. 1994. *Imagining the universe: A visual journey.* New York: Perigee Books/Berkeley Publishing.

Piaget, J. 1973. *The child and reality: Problems of genetic psychology.* New York: Grossman.

Posner, G. J., K. A. Strike, P. W. Hewson, and W. A. Gertzog. 1982. Accommodation of a scientific conception: Toward a theory of conceptual change. *Science Education* 66:211.

Rhoton, J., and P. Bowers, eds. 2003. *Science teacher retention: Mentoring and renewal.* Arlington, VA: NSTA Press.

Research Cited

Sae, A. S. 1996. *Chemical magic from the grocery store.* Dubuque, IA: Kendall/Hunt.

Sanders, W. L., and J. C. Rivers. 1996. *Cumulative and residual effects of teachers on future student academic achievement.* Knoxville: University of Tennessee, Value-Added Research and Assessment Center. *www.cgp.upenn.edu/pdf/Sanders_Rivers-TVASS_teacher%20effects.pdf*

Sarquis, M., J. P. Williams, and J. L. Sarquis. 1995. *Teaching chemistry with toys: Activities for grades K–9.* New York: McGraw-Hill/Terrific Science Press.

Schön, D. A. 1983. *The reflective practitioner: How professionals think in action.* New York: Basic Books.

Science Hobbyist: Amateur Science. Science myths in K–6 textbooks and popular culture. *www.amasci.com/miscon/miscon.html* (with extensive links to other sites)

Shulman, L. 1986. Those who understand: Knowledge growth in teaching. *Educational Researcher* 15 (2): 4–14.

Silver, B. L. 1998. *The ascent of science.* New York: Oxford University Press.

Singer, S. R., M. L. Hilton, and H. A. Schweingruber. 2006. *America's lab report: Investigations in high school science.* Washington, DC: National Academies Press. Executive Summary: *www.nap.edu/catalog/11311.html*

Spangler, S. 1995. *Taming the tornado tube: 50 weird and wacky things you can do with a tornado tube.* Englewood, CO: Wren Publishing.

Stamp, N., and T. O'Brien. 2005. GK-12 Partnership: A model to advance change in science education. *Bioscience* 55 (1): 70–77.

Stannard, C., T. O'Brien, and A. Telesca. 1994. STEP UP to networks for science teacher professional development. *Journal of Science Teacher Education* 5 (1): 30–35.

Stepans, J. 1996. *Targeting students' science misconceptions: Physical science activities using the conceptual change model.* 2nd ed. Riverview, FL: Idea Factory.

Stigler, J. W., and J. Hiebert. 1999. *The teaching gap: Best ideas from the world's teachers for improving education in the classroom.* New York: Simon and Schuster.

Stroop, J. R. 1935. Studies of interference in serial verbal reactions. *Journal of Experimental Psychology.* This article is available on the Classics in the History of Psychology website: *http//psychclassics.yorku.ca/Stroop*

Tauber, R. T., and C. Sargent Mester. 2007. *Acting lessons for teachers: Using performance skills in the classroom.* 2nd ed. Westport, CT: Praeger.

Taylor, B., J. Poth, and D. Portman. 1995. *Teaching physics with toys: Activities for grades K–9.* New York: McGraw-Hill/Terrific Science Press.

Texley, J., T. Kwan, and J. Summers. 2004. *Exploring safely: A guide for high school teachers.* Arlington, VA: NSTA Press.

Treagust, D. F., R. Duit, and B. J. Fraser, eds. 1996. *Improving teaching and learning in science and mathematics.* New York: Teachers College Press.

Vygotsky, L. S. 1978. *Mind in society.* Cambridge, MA: Harvard University Press.

White, R. T., and R. F. Gunstone. 1992. *Probing understanding.* London: Falmer.

Williams, P. 1999. *Exploring with Wisconsin fast plants.* Dubuque, IA: Kendall/Hunt.

Wolpert, L. 1992. *The unnatural nature of science.* Cambridge, MA: Harvard University Press.

Yager, R. E., ed. 2005. *Exemplary science: Best practices in professional development.* Arlington, VA: NSTA Press.

Youngson, R. 1998. *Scientific blunders: A brief history of how wrong scientists can sometimes be.* New York: Carroll and Graff Publishers.

Index

Index

B

Bacon, Francis, 244, 295
Bacon, Roger, 261
balloons
    in "air mass matters," 176
    in a bottle, 280–281
    hot air, 158, 159
    "needle through the balloon"
        activity, 211–219
    static electricity and, 201–209
    "two-balloon balancing act,"
        87–96
banking analogy of learning,
    259–265
BASF Superabsorbent Polymer
    Hygiene and Industrial
    Applications, 127
"batteries and bulbs," 97–108
    debriefing in, 102–105
    extensions for, 105–106
    internet connections with,
        106–107
    materials for, 99
    procedure for, 100–102
    science concepts in, 98
Becker Demonstrations, 159, 229,
    282, 328
Bell, Alexander Graham, 65, 75,
    80, 112
Benchmarks for Science Literacy, xx
    on "less is more" approach, 43
    on toys and engagement, 142
Ben Dosa, Hanina, 28
Bernoulli's principle, 147
Bernoulli's Principle Animation,
    148
bio-energy, 168
biological applications/analogies
    "brain-powered lightbulb"
        activity, 163–170
    "burning a candle at both ends"
        activity, 25–33
    "crystal heat" activity, 331–342
    "floating and sinking" activity,
        309–320
    "happy and sad bouncing balls"
        activity, 221–232

"identification detectives"
        activity, 73–83
"magnetic fields" activity,
    179–188
"mental puzzles, memory,
    mnemonics" activity,
    131–140
"needle through the balloon"
    activity, 211–219
optical illusions, 21, 47–61
perceptual paradoxes activities,
    37–46
"polarizing filters" activity,
    267–273
"sound tube toys" activity,
    141–151
"Stroop Effect" activity, 285–291
"super-absorbent polymers"
    activity, 119–129
"talking tapes," 109–117
"two-balloon balancing act,"
    87–96
"utensil music" activity, 63–71
biological cell membrane analogy,
    217
Biological Sciences Curriculum
    Study (BSCS), 356
Bizarre Stuff You Can Make in
    Your Kitchen, 339
black box experiments, 73–83
    "Cartesian diver," 321–329
    happy/sad balls, 228
Bloomfield, Louis A., 186
Bloom's Taxonomy, 247, 248
Boyle, Robert, 322
Boyle's law, 322
brain growth, 119–129
"brain-powered lightbulb" activity,
    163–170
    debriefing in, 167–168
    extensions for, 168–169
    internet connections for, 169
    materials for, 164–165
    procedure for, 166–167
    science concepts in, 164
brainstorming *vs.* brainwashing, 92
Brigham Young University Physics

Computer Resources, 272
Brilliant, Ashleigh, 183
Bruer, John T., 213, 218
Bruner, Jerome, 6
Bubbles, Babies and Biology, 94
buoyancy, 309–320
"burning a candle at both ends"
    activity, 25–33
    debriefing in, 29–31
    extensions for, 31–32
    internet connections for, 33
    materials for, 27
    procedure in, 28–29
    science concepts in, 27–28

C

Can Crush Demo, 176
cans, crushing, 175
"Cartesian diver," 321–329
    debriefing in, 325
    extensions for, 326–327
    internet connections for, 328–329
    materials for, 322–323
    procedure for, 323–324
    science concepts in, 322
CAST, 44, 229
catalysis, 331–342
change
    "happy and sad bouncing balls"
        and, 222–231
    law of, 206
    resistance to, 252
Changing Minds, 247
Chesterton, G. K., 155
Clarke, Arthur C., 21
cognition, xvi–xvii
    batteries and bulbs, 97–108
    brain growth and, 119–129
    "brain-powered lightbulb" and,
        163–170
    challenging learners' illusions in,
        259–265
    "cognitive inertia" activity,
        251–258
    conceptual filters and, 267–273
    "convection" activity and,
        153–162

Index

Dropper Popper toy, 228–229
dry ice, 337–338
Duit, R., 290, 299

E

economy, 346–347
"eddy currents" activity, 241–249
  debriefing in, 246
  extensions for, 247
  internet extensions for, 247–248
  materials for, 243
  procedure for, 244–246
  science concepts in, 242–243
Edible Candle, 31–32
Edison, Thomas Alva, 75, 80–81,
  110, 112, 191, 194–195
education concepts, 4–5
Education Trust, 18–19
effectiveness, 348–352
egg experiments, 297–298
Einstein, Albert, 165, 203, 224, 235,
  323
Elaborate phase, 359
e-learning, xxii
"electrical circuits: promoting
  learning communities"
  activity, 233–240
  debriefing in, 237–238
  extensions, 238–239
  internet connections for, 239–240
  materials for, 235
  procedure for, 236–237
  science concepts in, 234–235
"electric generators" activity,
  189–199
  debriefing in, 193–195
  extensions for, 195
  internet connections for, 195–196
  materials for, 191
  procedure for, 192–193
  science concepts in, 190
electricity
  "batteries and bulbs," 97–108
  "brain-powered lightbulb,"
    163–170
  "eddy currents" and, 241–249
  "electric generators" activity,

    189–199
  static, 201–209
Electronics for Kids, 106
emergent properties, 16
emotional engagement, xvii
"Empty the Bottle Race," 280
Encyclopedia of Science, 298
energy. *See* conservation of energy
Energy Balls, 233–240
engagement
  in 5E Teaching Cycle, 356–357
  fun first for, 74, 78
  learning communities and,
    233–240
  in "two-balloon balancing act,"
    92
  varying stimuli and, 141–151
enjoyable activities, 346–347
Escher, M.C., 12, 19, 20, 22, 60
ethics, "happy and sad bouncing
  balls" and, 221–232
Evaluate phase, 359–360
evidence, "happy and sad bouncing
  balls" and, 222–231
evolution
  brain, 120, 234
  "happy and sad bouncing balls"
    and, 222, 228
  of human senses, 39, 41, 45, 74,
    79
  novelty and, 142, 150, 151, 157
Explain phase, 358–359
Exploratorium, 59, 94, 138, 159
  on Cartesian diver, 328
  on tornado in a bottle, 306
Exploratorium Online
  on eddy currents, 247
  on magnetism, 187
  on motors, 196
  on polarization, 272
  on static electricity, 207
Exploratorium Snacks, 44
Exploratory Science Centre, 59
Explore phase, 357–358
eyedropper divers, 321–329
Eyetricks.com, 59

F

Faraday, Michael, 32, 180, 242, 244
"Fill That Bottle Race," 280
fire. *See* "burning a candle at both
  ends" activity
5E Teaching Cycle, 355–360
Flash Animations for Physics, 106,
  239
"floating and sinking" activity,
  309–320
  debriefing in, 314–315
  extensions for, 315–316
  internet connections for, 316–317
  materials for, 311
  procedure for, 312–314
  science concepts in, 310
Flubber (movie), 229
"Fortune Teller Miracle Fish"
  activity, 126
Franklin, Benjamin, 207
friction, 226
  "rattlebacks" and, 293–299
fruits, floating, 316
fun, engagement via, 74, 78, 202
  "floating and sinking" activity,
    309–320
  Stroop Effect and, 285–291

G

Galilei, Galileo, 122, 225, 253
Gardner, Martin, 21, 33
gas laws, 95–96
  air mass and, 171–177
  "invisible gases matter," 275–283
  kinetic molecular theory,
    211–219
  "tornado in a bottle" and,
    301–307
genetics, 114–115
genie in a bottle, 340
Gestalt psychology, 50–51, 59,
  183–184
Goethe, Johann Wolfgang von, 269
Gould, Stephen Jay, 4
gravity
  gravitational potential, 225, 226
  "rattlebacks" and, 293–299

Gregg, Alan, 52
Groovy Sounds, 115
guided discovery techniques, 132
Gutenberg, Johann, 237

H

Haldane, John, 90
hands-on explorations (HOEs), 55,
    104, 110, 123, xv
hand warmers, 338–339
"happy and sad bouncing balls"
    activity, 221–232
    debriefing in, 227–228
    extensions for, 228–229
    internet connections for, 229–230
    materials for, 223
    procedure for, 224–227
    science concepts in, 222–223
Hawking, Stephen, 4
hearing experiments, 63–71
    "talking tapes," 109–117
heat
    "burning a candle at both ends"
        activity, 25–33
    "crystal heat," 331–342
Hery, Joseph, 242
Highet, Gilbert, 17
history and philosophy of science
    (HPS), xix
History Guide, The, 44
HomeHarvest Garden Supply's Soil
    Moist, 127
hot air balloons, 158, 159
Howard Hughes Medical Institute,
    44, 60
*How Everything Works: Making
    Physics Out of the Ordinary*
    (Bloomfield), 186
HowStuffWorks, 33, 44, 69, 106,
    115
    on convection, 159
    on disposable diapers, 127
    on electricity, 196
    on magnetism, 187
    on memory, 138
    on polarization, 272
    on static electricity, 207

    on submarines, 328
    on vacuum cleaners, 148
Hume, David, 165
Huxley, Thomas Henry, 155, 203
HyperPhysics, 106, 176, 247, 328

I

"identification detectives" activity,
    73–83
    debriefing in, 78
    extensions for, 78–81
    internet connections with, 81–82
    materials for, 74–75
    procedure for, 76–77
    science concepts in, 74
Illusions-Optical.com, 60, 230
Illusion Works, 60, 263
implementation dip, 268
Indestructables, 187
inertia, 172. *See also* motion, laws of
    "air mass matters" activity,
        171–177
    "cognitive inertia" activity and,
        251–258
    "rattlebacks" activity and,
        293–299
inertia, cognitive. *See* cognitive
    inertia
inquiry
    analogies for, 3–15, xv
    inspiring, 174–175
    "mental puzzles, memory,
        mnemonics" activity,
        131–140
    Möbius strip activity, 15–23
inservice learning, xiii–xiv
Institute for Human and Machine
    Cognition (IHMC), 187
Instructables, 340
Instructional Design Models, 12
instruction model, 355–360
integrated instructional unit design,
    355–360
intelligence theories, 44, 74, 229,
    230
International Mind Brain and
    Education Society, 44, 81,
    169, 187

internet connections, xxi–xxii
    for "air mass matters," 176–177
    for analogies, 12–13
    with "batteries and bulbs,"
        106–107
    for "brain-powered light bulb,"
        169
    for "Cartesian diver," 328–329
    for "cognitive inertia" activity,
        256–257
    for "convection" activity,
        159–160
    for "crystal heat" activity,
        339–340
    for "eddy currents" activity,
        247–248
    for "electrical circuits: promoting
        learning communities"
        activity, 239–240
    for "electric generators,"
        195–196
    for "floating and sinking"
        activity, 316–317
    for "happy and sad bouncing
        balls," 229–230
    for "identification detectives,"
        81–82
    for "invisible gases matter"
        activity, 282
    for "magnetic fields," 187–188
    for "mental puzzles, memory,
        mnemonics," 137–138
Möbius strip, 22
    for "needle through the balloon,"
        218
    for optical illusions, 59–60
    for "optics and mirrors" activity,
        263–264
    for perceptual paradoxes, 43–44
    for "rattlebacks" activity,
        298–299
    for "sound tube toys," 148
    for "static electricity," 207–208
    for "Stroop Effect" activity,
        290–291
    for "super-absorbent polymers,"
        127

Index

Index

Index

on bouncing balls, 230
on density and buoyancy, 317
on eddy currents, 248
on electricity and magnetism, 196
on heat and fluids, 328
on inertia, 257
on mechanics, 298
on polarization, 272
on static electricity, 207
University of Minnesota, 340
University of Victoria, 248
University of Virginia Phun Physics Show, 208, 230
on inertia, 257
University of Virginia Physics Department, 44, 70, 107, 148, 160, 176, 188
on air properties, 282
on bouncing balls, 230
on eddy currents, 248
on electrical circuits, 239
on electromagnets, 196
on inertia, 257
on polarization, 272
on static electricity, 208
University of Wisconsin-Madison Physics Lecture Demonstrations, 257
"utensil music" activity, 63–71
debriefing for, 66–67
extensions for, 67–69
internet connections for, 69–70
materials for, 64
procedure for, 65–66
science concepts in, 64

V

Virginia Tech Physics Lecture Demo, 82
Virtual Circuit Simulator/Lab, 240
Virtual Voltage Circuit Simulator, 107
Visible Thinking, 188
Vision Science, 60

visual participatory analogies. *See* analogies
von Humboldt, Alexander, 122
von Liebig, Justus, 277

W

wait time, 242–243
Wake Forest University Physics Department, 70, 177, 188, 208, 257, 328
Ward's Natural Science Co., 82
"Waste Paint Hardener" activity, 126
Watersorb, 127
websites. *See* Internet connections
weight experiments, 41, 43
Whelmers, 33, 70, 148, 160, 177, 208, 329
Wikipedia
on analogies, 13
on attention, 291
on batteries and bulbs, 107
on bouyancy, 317, 329
on catalysis, 340
on cognitive biases, 299
on cognitive connections, 188
on cognitive load theory, 44, 177
on cognitive neuroscience, 169
on constructivism, 264
on convection, 160
on eddy currents, 248
on electrical generators, 196
on inertia, 257
on LaPlace's law, 94
on membranes, 218
on memory, 138
on Möbius strips, 22
on optical illusions, 60
on polarization, 272
on rattlebacks, 298
on sound, 70, 148
on sound recording, 115
on static electricity, 208
on superabsorbent polymers, 127
on vortex, 306

Wilson, Edward O., 4
Wilson, Erasmus, 165
Wolfram MathWorld, 22
Woodrow Wilson Leadership Program in Chemistry, 340
wow-and-wonder-before-words approach, 174

Y

Yeats, William Butler, 28

Z

zone of proximal development (ZPD), 172